a new introduction to
GEOGRAPHY
for OCR GCSE Specification A

Written by:

David Gardner, Greg Hart and Nic Howes

With contributions from:

John Belfield, John Pallister and Brian Smith

Hodder & Stoughton
A MEMBER OF THE HODDER HEADLINE GROUP

Acknowledgements

Photograph credits
The authors and publisher thank the following for permission to reproduce their photographs in this book.

Fig 1.11 (top) George Hall/Corbis; Fig 1.11 (btm) Roger Ressmeyer/Corbis; Fig 1.20 (top) John Meeham/Science Photo Library, (centre 3) Stewart Lowther/Science Photo Library, (btm) David Weintraub/Science Photo Library; Fig 1.22 Gary Rosenquist/USGS; Figs 1.23, 1.24 1.25 Lyn Topinka/USGS; Fig 1.27 US Geological Center; Fig 1.30 Thomas Raupach/Still Pictures; Fig 1.31 Assoociated Press AP; Figs 1.36, 1.37 1.38 Nic Howes; Fig .l.43 Skyscan Photolibrary/William Cross; Figs 1.46, 1.48 Popperfoto/Reuter; Figs 1.52, 1.53 Associated Press AP; Fig 1.54 Jim Holmes/Panos Pictures; Figs 1.57, 1.59, 1.67 Skyscan Photolibrary; Fig 1.69 Trackair Aerial Surveys; Figs 1.71, 1.72 Ross-Parry Picture Agency; Fig 1.78 Hull Daily Mail Publications Ltd; Figs 1.79, 1.81, 1.89, 1.91, 1.92, 1.93, 1.95 John Pallister; p62 SJ Sibley; Fig 2.29 Annie Griffiths Belt/CORBIS; Fig 2.30 Robert Harding; Fig 2.37 Associated Press AP; Figs 2.40, 2.41, 2.42, 2.43 Keith Flinders; Fig 2.44 Courtesy of the City and County of Swansea; Figs 2.45, 2.46, 2.48, 2.49, 2.50 Keith Flinders, Fig 2.52 Val Flinders, Figs 2.54, 2.56, 2.57, 2.58, 2.59 Keith Flinders; Figs 2.61, 2.65, 2.66, 2.68, 2.71 John Pallister; pp126-127 Keith Flinders; Fig 3.11 Jorgen Schytte/Still Pictures; Figs 3.25, 3.27 David Gardner; Fig 3.29 Joerg Boethling/Still Pictures; Fig 3.30 Emma Lee/Life File; Fig 3.35, 3.37 Sealand Aerial Photography; Fig 3.39 Keith Flindes, Fig3.40 A.J. Williamson, Fig 3.41 Keith Flinders, Fig 3.44 Stephen Hardy, Fig 3.45 Judith Capton, Figs 3.52, 3.51, 3.57 Keith Flinders; Fig 3.61 Richard Bird Photography; Figs 3.62, 3.63 Keith Flinders; Figs 3,65, 3.66, 3.67 John Pallister; pp 178-179 Keith Flinders; Figs 4.11, 4.13, 4.14, Sue Cunningham/SCP;Fig 4.12 Andrew Ward/Life File; Fig 4.7Mark Edwards/Still Pictures; Fig 4.18 Nigel Dickinson/Still Pictures; Figs 4.20, 4.22, 4.23 Keith Flinders; Figs 4.25, 4.26 Associated Press EFE; Fig 4.27 Associated Press AP; Fig 4.28, 4.29 ©2000 German Aerospace Center (DLR); Fig 4.32 University of Dundee; Fig 4.31Chinch Gryniewicz; Ecoscene/CORBIS; Fig 4.33 Bryan Pickering, Eye Ubiquitous/CORBIS; Fig 4.35 Ecoscene/Leeney; Fig 4.37 Ecoscene/Schaffer; Fig 4.42 Associated Press/Greenpeace via PA; Fig 4.44 Andrew Testa/The Observer; Fig 4.48 John Pallister; p215 Greenpeace/Morgan; pp 178, 220 Brian Smith, pp221, 224-225, 226 Keith Flinders; p 227 Emma Flinders, p237 S.J. Sibley.

Other credits
The authors and publisher are also grateful to the following for permission to reproduce their copyright materials either in their original form or in a form adapted for the purposes of this book:

Fig 1.11 (home page) Mount St. Helens Tours Inc; Fig 1.11 based on a graphic © Times Newspapers Limited, 19th October 1989; Fig 1.12 San Francisco Chronicle; Fig 1.13 Los Angeles Times Syndicate; Fig 1.14 San Francisco Chronicle; Fig 1.17 Los Angeles Times Syndicate; Fig 1.19 USGS; Fig 1.26 Haraldur Sigurdsson; Fig 1.29 © Times Newspapers Limited, 8th December 1998; Fig 1.77 based on a graphic © Telegrpah Group Ltd; Fig 2.31 © Based on Bartholomew Digital Database. Reproduced by permission of HarperCollins Publishers; Fig 2.32 Longman Geography for GCSE, p120, Longman; Fig 3.9 Hamish McRae/ The Independent Syndication; Fig 3.17 © Times Newspapers Limited, 25th October 1998; p147 FAO (Food and Agriculture Organization of the United Nations);p150 BBC Online News; Fig 3.32 © Times Newspapers Limited, 13th November 1994; Fig 3.33 cartoon by Austin, eds. J. Porritt, and R. Maynard , *Earth Mirth*, 1986, Friends of the Earth; Fig 3.59 Sheffield Telegraph; p 187 BBC Online News; Fig 4.19 Lagamar Expeditions; p191 BBC Online News; Fig 4.38 © Telegraph Group Ltd, 2000; Fig 4.38 (inset) Climatic Research Unit, University of East Anglia; Fig 4.41 US Department of Commerce: National Oceanic and Atmospheric Administration/The Telegraph Group Limited/Richard Burgess; Fig. 1 p212, Oliver & Boyd, GCSE Copymasters, Adddison Wesley Longman Ltd; Fig. 2 p213 Greenpeace UK; Fig 7 p237 *Energy Resources for a Changing World*, JE Allan, 1992, Cambridge University Press.
Maps reproduced from Ordnance Survey mapping with permission of the Controller of Her Majesty's Stationery Office © Crown copyright Licence Number 399450: OS Landranger 1:50 000 series: Fig 1.63, Fig 2.55, p 121, Fig 3.64; OS Explorer 1:25 000 series: Fig 1.43; OS Outdoor Leisure 1:25 00 series: p230.

Every effort has been made to contact copyright holders and to obtain permission to reproduce copyright materials. If any have been overlooked, the publishers will make the appropriate arrangements at the first opportunity.

Note about the Internet links in the book

The user should be aware that URLs or web addresses change regularly. Every effort has been made to ensure the accuracy of the URLs provided in this book on going to press. It is inevitable, however, that some will change. It is sometimes possible to find a relocated web page, by just typing in the address of the home page for a website in the URL window of your web browser.

Words in bold

Words that appear in bold are key syllabus words that are either explained in the text, or in the glossary on page 239.

Ordnance Survey Key

A key to the OS Landranger maps can be found on page 123

Orders: please contact Bookpoint Ltd, 130 Milton Park, Abingdon, Oxon OX14 4SB.
Telephone: (44) 01235 827720, Fax: (44) 01235 400454. Lines are open from 9.00 – 6.00,
Monday to Saturday, with a 24 hour message answering service. Email address: orders@bookpoint.co.uk

British Library Cataloguing in Publication Data
A catalogue record for this title is available from The British Library

ISBN 0 340 74707 2

First published 2001
Impression number   5   4   3
Year   2005 2004 2003 2002 2001

Typeset by Liz Rowe.
Printed in Italy for Hodder & Stoughton Educational, a division of Hodder Headline Plc, 338 Euston Road, London NW1 3BH by Printer Trento.

Contents

Liverpool
Community
College

UNIT 4   PEOPLE AND THE ENVIRONMENT

one

People and the Physical World

Plate tectonics

WHAT CAUSES VOLCANOES AND EARTHQUAKES?

To understand why volcanoes and earthquakes happen, we need to understand what is happening beneath the earth's surface.

Little is known about inside the earth. We do not have access to it. Most of what we do know comes from studying the way earthquake waves pass through the earth. It consists of several concentric layers made of different rocks and substances that stretch more than 6 000 km from the inner core to the surface crust. Figure 1.1 is a cross-section through the interior of the earth. It shows that there are three main parts. They are the **core**, **mantle** and the **crust**. Each of these layers is different, due to density, rock type and temperature.

The core is the central part of the earth, where temperatures and pressure are enormous. It is partly solid, the inner core, and partly semi-liquid, the outer core.

The mantle is made up of rocks and is sandwiched between the core and the crust. It is the thickest layer, occupying over 80% of the earth's volume. It consists of mainly solid rocks, but the upper mantle is actually a layer of semi-liquid molten rock called magma. This magma flows slowly underneath the crust, moved by convection currents.

The crust is the relatively thin outermost layer of rocks that 'float' on the mantle. It is solid rock. The

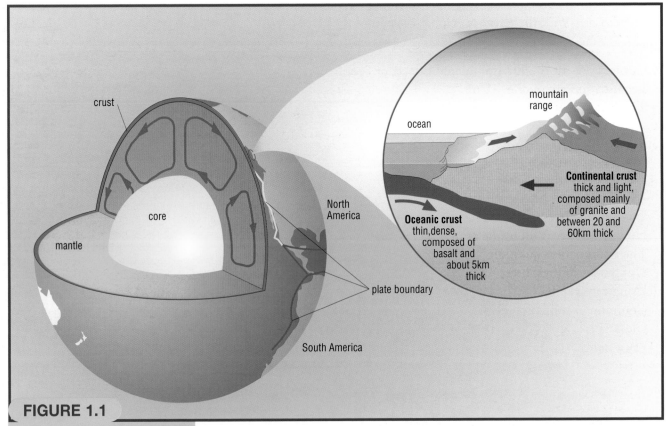

FIGURE 1.1

A cross-section through the earth

crust is not one continuous piece. It is broken into several large and other smaller sections, known as **plates** (or tectonic plates). These plates fit together like pieces of a jigsaw puzzle. They float like rafts on the molten rock of the mantle. Some plates move a few centimetres every year, at about the same speed as a fingernail grows. The theory that describes their movement is called plate tectonics. The place where two plates meet is called a plate boundary (or margin), and it is here that most of the world's earthquakes occur and volcanoes erupt.

There are two types of crust, continental and oceanic crust. Continental crust is older and lighter. It cannot sink and is permanent. Oceanic crust is younger and heavier. It can sink and is constantly being destroyed and replaced.

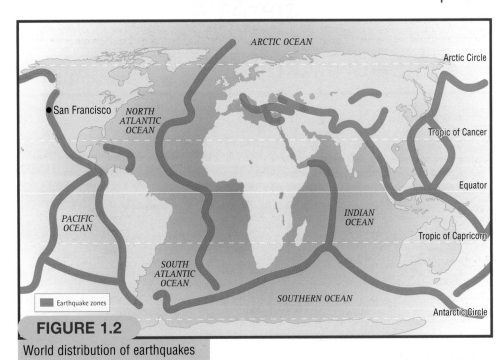

FIGURE 1.2

World distribution of earthquakes

FIGURE 1.3

World distribution of volcanoes

TASKS

1 What is a plate?

2 With the aid of a diagram, explain why plates move.

3 Figures 1.2 and 1.3 show the world's distribution of earthquakes and volcanoes. They both occur in long narrow belts around the world.
   a Using an atlas, describe the patterns of earthquake and volcanic activity.
   b Give examples of continents and oceans where they occur.

4 a Add to an outline map of the world the distribution of earthquakes and volcanoes.
   b Research with an atlas, reference books, CD-ROM or websites the location and date of ten major earthquakes and ten major volcanic eruptions. Mark and name them on your outline map, adding the year in brackets.

PLATE BOUNDARIES

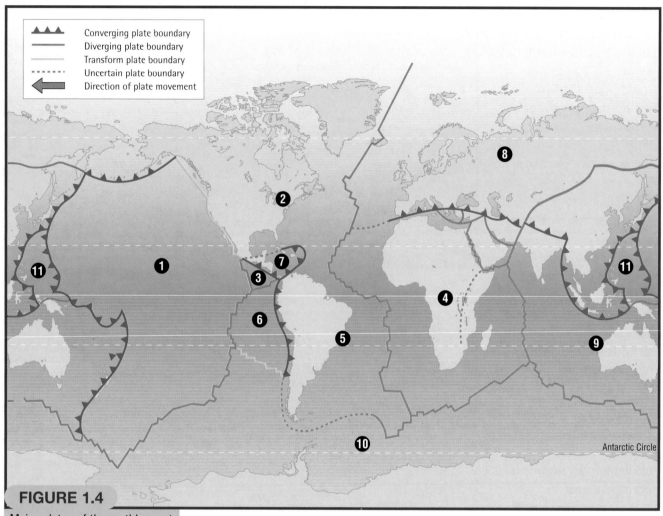

FIGURE 1.4

Major plates of the earth's crust

Earthquakes and volcanoes often occur in similar places around the world. These locations are at or near the edges or boundaries of plates. The distribution of plates and their boundaries are shown on Figure 1.4; compare them to Figures 1.2 and 1.3. The key to the map shows that there are three types of plate boundary – **convergent**, **divergent** and **transform**.

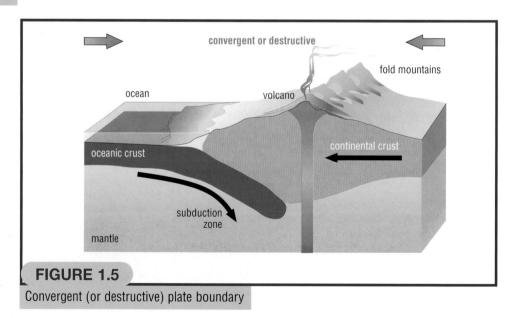

FIGURE 1.5

Convergent (or destructive) plate boundary

At convergent or destructive boundaries, one plate collides with another. When oceanic crust collides with continental crust, the oceanic crust is forced downwards into the mantle and destroyed. This is called a subduction zone. Where two plates made up of continental crust collide, they are pushed upwards forming fold mountains such as the Himalayas and Andes.

Along divergent or constructive boundaries, plates are forced apart by movement of the magma in the mantle and new crust is formed in between.

Transform or passive boundaries are where two plates move horizontally past each other.

FIGURE 1.7

Transform plate boundary

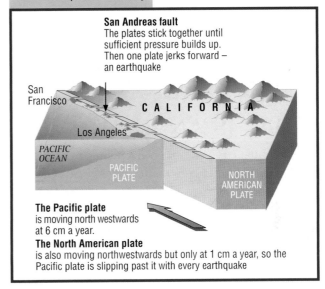

San Andreas fault
The plates stick together until sufficient pressure builds up. Then one plate jerks forward – an earthquake

San Francisco

CALIFORNIA

Los Angeles

PACIFIC OCEAN

PACIFIC PLATE

NORTH AMERICAN PLATE

The Pacific plate
is moving north westwards at 6 cm a year.
The North American plate
is also moving northwestwards but only at 1 cm a year, so the Pacific plate is slipping past it with every earthquake

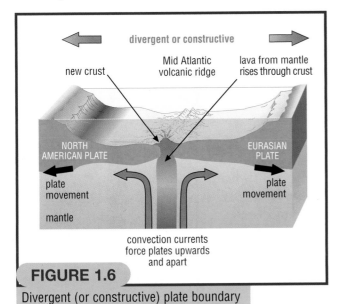

divergent or constructive

Mid Atlantic volcanic ridge

lava from mantle rises through crust

new crust

NORTH AMERICAN PLATE

EURASIAN PLATE

plate movement

plate movement

mantle

convection currents force plates upwards and apart

FIGURE 1.6

Divergent (or constructive) plate boundary

TASKS

1 On your outline map of the world, use different colours to show the three types of plate boundary.

2 On Figure 1.4, the names of the plates have been labelled 1–11. Using an atlas and the list of names given below, label the names of the plates on your map.

Eurasian; North American; South American; African; Pacific; Indo-Australian; Nazca; Cocos; Antarctic; Philippine; Caribbean.

3 What type of plate boundary (convergent, divergent or transform) exists between the following plates:
   a North American and Eurasian
   b North American and Pacific
   c Nazca and South American
   d Indo-Australian and Eurasian
   e African and South American

4 Write an explanation of the following terms, and use diagrams to help explain features:
   a convergent plate boundary
   b divergent plate boundary
   c transform plate boundary
   d subduction zone
   e fold mountains
   f magma
   g crust

5 Why do earthquakes occur along:
   a subduction zones?
   b transform plate boundaries?

6 Why do volcanoes occur along:
   a subduction zones?
   b divergent plate boundaries?

7 You can find out further information about plate boundaries by conducting research using CD-ROMs and the Internet. You could create your own report about plates by copying and pasting diagrams, text and photographs into a word processing file, where you could label and edited the data.

The following sources have useful information:
CD-ROMs, such as *Encarta Encyclopaedia*; *Physical World*; *Geodome Landforms*
Internet

Understanding plate motions [This Dynamic Earth, USGS]
http://pubs.usgs.gov/publications/text/understanding.html

Cascades Volcano Observatory HOME PAGE

http://vulcan.wr.usgs.gov/home.html

Select Menus of interest; glossary of Hazards, Features, and Terminology and then select plate tectonics.

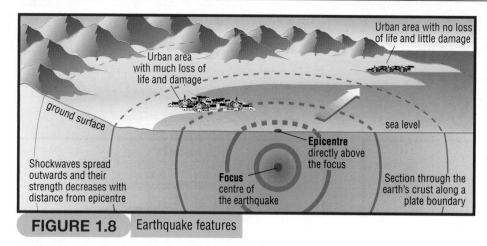

FIGURE 1.8 Earthquake features

WHAT IS AN EARTHQUAKE?

An earthquake is a sudden movement in the earth's crust. It is the result of a sudden release of energy that causes shock or seismic waves. The point where the pressure is released, the centre of the earthquake, is called the **focus**. (Figure 1.8) The **epicentre** is the point on the earth's surface immediately above the focus. The shock waves radiate outwards from the focus. It is at or near the epicentre where most damage is done; as the waves spread out they decrease in strength until they become harmless, like ripples in a pond.

How are earthquakes measured?

The strength of an earthquake is recorded using an instrument called a **seismograph**. The strength of the earthquake is measured on the **Richter** scale and given a value between 1 and 10. The scale is logarithmic, so each point is ten times greater than the previous one. An earthquake at 7 on the scale is 10 times greater than one measuring 6, and 100 times greater than one measuring 5.

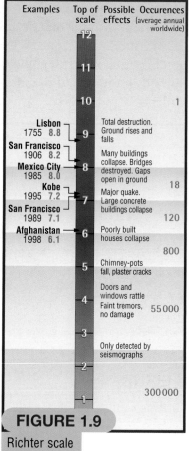

FIGURE 1.9

Richter scale

Year	Place	No. of deaths	Richter scale
1906	San Francisco, USA	450	8.2
1923	Tokyo, Japan	156 000	7.9
1964	Alaska, USA	131	8.3
1970	Peru	66 000	7.8
1971	San Fernando, USA	65	6.5
1976	Tangshan, China	240 000	7.6
1985	Mexico City	9 500	8.1
1988	Armenia	25 000	6.9
1989	San Francisco, USA	67	7.1
1990	Iran	50 000	7.3
1993	Latur, India	10 000	6.5
1994	Los Angeles, USA	57	6.6
1995	Kobe, Japan	5 477	7.2
1998	Afghanistan	4 000	6.1
1999	Turkey	3 000	7.4

Figure 1.10

TASKS

1 Write an explanation of the following terms; use diagrams to help:
   a earthquake   c focus
   b epicentre   d seismic waves
   Draw a diagram to show that you have understood the terms.

2 a With the aid of an atlas, mark the location of each earthquake shown in Figure 1.10, name the location and record the date and scale of the quake.
   b Compare the map with your plate boundary map. Describe the distribution of the earthquakes in relation to the three types of plate boundary – divergent, convergent and transform.
   c Plot a scattergraph of the death toll for each earthquake shown in the table (y axis) against its strength (x axis).
   d Describe the relationships you find in your graph. How strong is the link between earthquake intensity and the number of deaths caused? Suggest reasons for this.

3 Refer to the Richter scale diagram (Figure 1.9). How many times stronger was the 1906 San Francisco earthquake than those in:
   a San Francisco in 1989
   b Afghanistan in 1998.

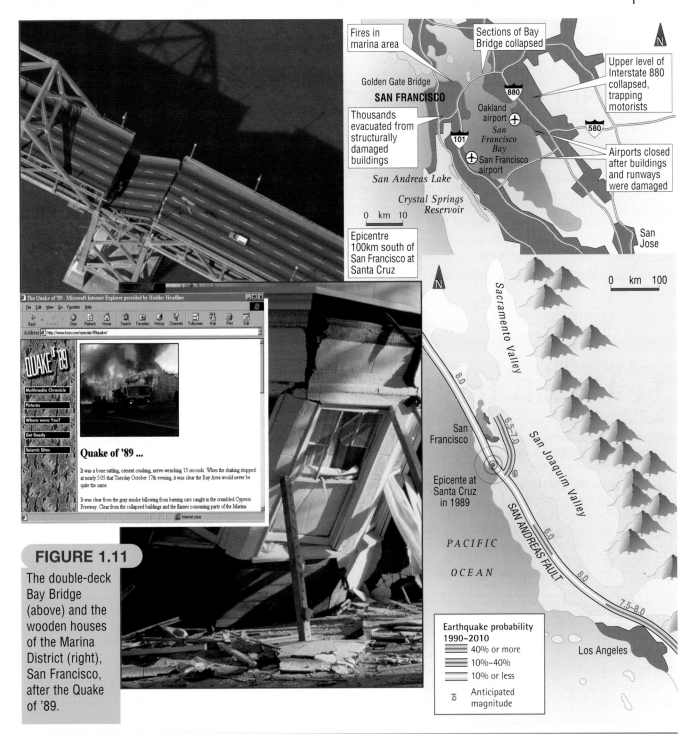

Fires in marina area

Sections of Bay Bridge collapsed

Upper level of Interstate 880 collapsed, trapping motorists

Golden Gate Bridge

SAN FRANCISCO

Thousands evacuated from structurally damaged buildings

Oakland airport

San Francisco Bay

San Francisco airport

101

880

580

Airports closed after buildings and runways were damaged

San Andreas Lake

Crystal Springs Reservoir

0   km   10

San Jose

Epicentre 100km south of San Francisco at Santa Cruz

The Quake of '89 - Microsoft Internet Explorer provided by Hodder Headline

File Edit View Go Favorites Help

Back Forward Stop Refresh Home Search Favorites History Channels Fullscreen Mail Print Edit

Address http://www.kron.com/specials/89quake/

QUAKE OF '89

Multimedia Chronicle

Pictures

Where were You?

Get Ready

Seismic Sites

Quake of '89 ...

It was a bone rattling, cement crushing, nerve-wracking 15 seconds. When the shaking stopped at nearly 5.05 that Tuesday October 17th evening, it was clear the Bay Area would never be quite the same.

It was clear from the gray smoke billowing from burning cars caught in the crumbled Cypress Freeway. Clear from the collapsed buildings and the flames consuming parts of the Marina.

Internet zone

Sacramento Valley

0   km   100

8.0

San Francisco

6.5–7.0

San Joaquim Valley

San Andreas Fault

Epicente at Santa Cruz in 1989

6.0

PACIFIC

OCEAN

8.0

7.5–8.0

Los Angeles

Earthquake probability 1990–2010

40% or more

10%–40%

10% or less

Anticipated magnitude

FIGURE 1.11

The double-deck Bay Bridge (above) and the wooden houses of the Marina District (right), San Francisco, after the Quake of '89.

TASKS

4 The resources on this page give information about the earthquake which struck San Francisco in 1989. This was a moderate quake whose epicentre, Loma Prieta, was located more than 80 km south of San Francisco and Oakland in the Santa Cruz Mountains on the San Andreas Fault. Your task is to write an investigation of the earthquake using the resources on this page, together with any additional information you can obtain from CD-ROMs or the Internet (such as www.sfmuseum.org/). You could present your findings in the form of a word processed or desktop published report.

Try to use the following headings:

A description of the events – try to include quotes from eye-witnesses, fact and figures and photographs.

Identify **the causes of the earthquake**, explaining your findings with the aid of diagrams.

A description of **the effects of the earthquake**.

WHY DO MANY PEOPLE CONTINUE TO LIVE AT RISK FROM EARTHQUAKES?

Earthquakes hit major centres of population often, yet millions of people continue to live in areas prone to such disasters. San Francisco was almost completely destroyed by an earthquake in 1906. The city was rebuilt on the same site, even though people knew that further earthquakes were likely to happen there.

It is not possible to prevent earthquakes. They usually occur without any warning. All people can try to do is:

▶ Prepare for them to reduce the damage, and ensure that the emergency services will be available.

▶ Try to predict where and when an earthquake will happen, but this is extremely difficult. As a result, scientists and engineers have concentrated on being prepared rather than prediction.

Earthquake proof buildings

In major MEDC cities which are prone to earthquakes, such as San Francisco and Tokyo, new

---

San Francisco Chronicle   **Thursday October 17 1996**

Few ready for the next big one, poll finds

Charles Petit

If the Loma Prieta earthquake was a wake-up call to Northern California, most people seem to have gone back to sleep. Even the government is cutting funds for earthquake studies.

A poll being released today by the American Red Cross and Pacific Gas and Electric Co. shows that 82% of Northern and Central California residents say that they are "not very well prepared" for the next big disaster, and only 25% believe a major disaster is likely in the next 10 years. In fact, some scientists have estimated the risk of an earthquake in the immediate Bay area as high as 90% during the next 30 years.

The Red Cross survey found a variety of reasons people are not ready for a big quake. They include a feeling that its not worth their time (19%); ignorance of what to do (17%); laziness (11%); no time to do it (11%); no money to do it (10%); and belief no disaster will happen (6%), others (26%)

Those under the age of 35, who presumably have experienced fewer disasters are the least prepared, with 95% saying they have done little or nothing to get ready.

Figure 1.12

As the population of Turkey has increased, many people have left the countryside believing they would have a better life in a city. The numbers living in Istanbul soared nearly 12-fold since 1945, from 850 000 to 10 million in 1999. This area was identified as the likely site of a major earthquake 15 years before the disasterous earthquake of August 1999. Despite this, poor quality homes were thrown up in parts of Istanbul and its surroundings to house the booming population. This increased the risk of death and injury during the earthquake.

People are prepared to live with the risk because the advantages of job opportunities and a higher standard of living outweigh the dangers they perceive. After all, it might not happen in their lifetime.

buildings are designed to withstand earthquakes (Figure 1.13). In the Loma Prieta earthquake of 1989, 60 miles away in downtown San Francisco, the occupants of the Transamerica Pyramid were unnerved as the 49-storey office building shook for more than a minute. US Geological Survey (USGS) instruments, installed years earlier, showed that the top floor swayed more than 30 cm from side to side. However, no one was seriously injured and the Transamerica Pyramid was not damaged. This famous San Francisco landmark had been designed to withstand even greater earthquake stresses, and its design worked as planned during the earthquake. As scientists learn more about ground motion during earthquakes and structural engineers use this information to design stronger buildings, loss of life and property can be reduced.

Thursday, August 19, 1999 — Los Angeles Times

L.A. Subject to Same Jolt but is Better Built

By ROBERT LEE HOTZ, Times Science Writer

On any given day in metropolitan Los Angeles, there are half a dozen faults capable of creating an earthquake like the explosive magnitude 7.4 temor that struck Turkey before dawn Tuesday, killing thousands. But a similar quake here would have far less devastating results, earthquake experts said Wednesday.

The reason is less a matter of seismology or fault planes than of bricks and mortar. When it comes to earthquake engineering, California has perhaps the world's strictest building codes, which have been updated regularly over the decades as experience with urban earthquakes has grown. More important, those codes have been more strictly enforced, several experts said.

In Southern California, more people also live in flexible, wood-frame homes, not the concrete high-rise apartments common in urban Turkey. The quality of housing construction — also a major cause of casualties in a severe 1995 quake in Kobe, Japan — may have been the key factor in the high death toll from the most recent urban earthquake disaster. "We are better prepared because we are better off economically and have more funds to put into each building," said UCLA geophysicist David Jackson. ■

Figure 1.13

READY OR NOT

Rick DelVecchio, San Francisco Chronicle Sunday, January 10, 1999

Bay Area earthquake researchers are racing the clock, gathering information they hope will cut down the damage and suffering when a big one rips the East Bay's Hayward Fault

When it comes to earthquakes, as with heart attacks, there are two ways to pay. Pay now for preventive surgery, or do nothing and pay a much higher price later when the attack hits. To many people, earthquakes don't seem real. It's difficult to grasp their enormity, and the big ones are so infrequent they don't make much of a dent on the collective memory.

Recent quakes in the Bay Area, have, for most people, been like rolling on a water bed. But a truly major quake will be a different order of experience. Thousands of people, if not millions, will feel as if they have been seized by the collar and tossed into the air. A half-minute of shaking will cause death, injury and widespread economic loss, but its effects won't be spread out evenly across the region. Just as geographical regions have microclimates, it turns out that earthquakes have cool and hot spots.

Both close to the epicenter and far from it, some spots will be hit far worse than others, and scientists are in a hurry to try to figure out the likely patterns.

Within a year, the U.S. Geological Survey hopes to publish new seismic geology maps as detailed as the contour maps that hikers use. Their maps will also show new data on the risk of quake-related landslides in the East Bay hills.

Such maps could influence economic decisions related to quake preparedness, such as insurance rates and building plans. They would also help in designing and upgrading buildings to withstand different levels of damage. A city with many thousands of structures built or shored up according to these principles would be able to snap back more quickly from disaster.

Figure 1.14

One of the best ways to reduce the impact of an earthquake is to avoid building in areas of high risk. During earthquakes, shock waves cause water to rise to the surface turning clay into mud. This process is called **liquefaction**. Any buildings built on clay lose their foundations and sink into the mud as they collapse. This happened during the Loma Prieta earthquake in San Francisco's Marina District, (Figure 1.11) which was built on reclaimed land. These were the only buildings to collapse during that earthquake.

Improved preparation

In many places prone to earthquakes, people try to prepare themselves. In Japan, 1 September is Disaster Day, the anniversary of the devastating Tokyo earthquake of 1923, which killed 156 000 people. The day is a public holiday, when Japanese people practise earthquake drill. They simulate what an earthquake would be like, how to protect themselves during an earthquake, how to fight fires and give first aid. In the USA however, government agencies and the media provide extensive guidance about what to do before, during and after an earthquake. This information is available for people on the Internet: Earthquake Information from the USGS http://quake.usgs.gov/ This site provides a wide range of information about earthquakes including advice on how to prepare for an event.

FIGURE 1.15 Earthquake survival kit

Shelf Life

- 3 months
- 6 months
- 1 year

Top

- Batteries, with tester
  Flashlight
  Portable radio
  First aid kit

Middle

- Food and water for pets, manual can opener, dry food (pasta, rice)
- Instant food, water, purification tablets
- Canned food

Bottom

- Blanket
  Tarpaulin
  Extra clothing, shoes
  Premoistened towelettes
  Items for personal hygiene: toilet tissue and heavy-duty plastic bags for disposal

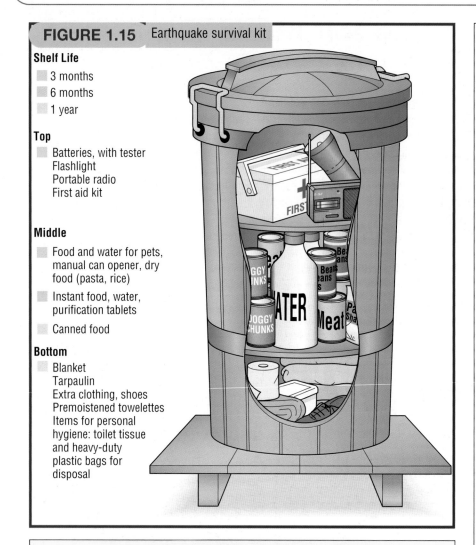

EARTHQUAKE KIT

A Guide to Updating Emergency Supplies
Los Angeles Times Earthquake Safety Guide

Earthquake preparedness consists of a series of precautions in the form of survival kits, which usually can be started or replenished with supplies already on the shelf. Encourage the folks next door to update their emergency supplies too.

Recommendations vary as to the appropriate amount of emergency food and water to store. Many experts advise one gallon of water per person per day for three to seven days. Also store enough food for the same period of time.

Use a large container such as a footlocker or 30-gallon trash can, and label each food and water item with the date of purchase or the last date it should be used. Place the container in a cool, dark place, such as a garage, on something to raise it off the ground.

MARK HAFER / Los Angeles Times

Figure 1.16

Los Angeles Times Earthquake Safety Guide

What to do during an earthquake

1 If you are INDOORS – STAY THERE! (Get under a desk or table and hang on to it, or move into a hallway or get against an inside wall. STAY CLEAR of windows, fireplaces, and heavy furniture or appliances. GET OUT of the kitchen, which is a dangerous place (things can fall on you). DON'T run downstairs or rush outside while the building is shaking or while there is danger of falling and hurting yourself or being hit by falling glass or debris.

2 If you are OUTSIDE – get into the OPEN, away from buildings, power lines, chimneys, and anything else that might fall on you.

3 If you are DRIVING – stop, but carefully. Move your car as far out of traffic as possible. DO NOT stop on or under a bridge or overpass or under trees, light posts, power lines, or signs. STAY INSIDE your car until the shaking stops. When you RESUME driving watch for breaks in the pavement, fallen rocks, and bumps in the road at bridge approaches.

4 If you are in a MOUNTAINOUS AREA – watch out for falling rock, landslides, trees, and other debris that could be loosened by quakes.

Figure 1.17

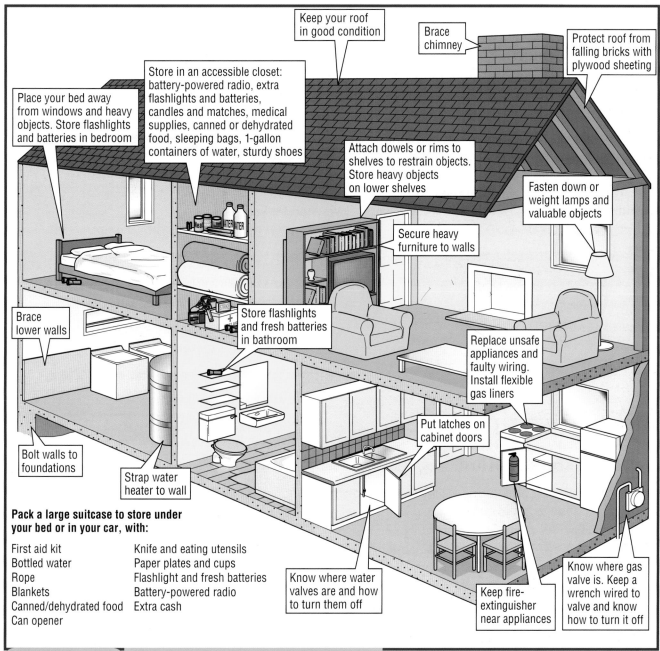

FIGURE 1.18 Preparing your house to survive an earthquake

TASKS

1 a Figure 1.12 is an extract from the San Francisco Chronicle in 1996. Read it carefully, then draw a pie chart to show the reasons why people are not ready for a major earthquake.
   b Explain what it shows.

2 Why do many people continue to live in cities which are at risk from earthquakes?

3 Why have scientists and engineers concentrated on earthquake preparedness rather than prediction?

4 Figure 1.13 is an extract from the Los Angeles Times in 1999. Read it carefully, then explain why earthquakes are more likely to be a disaster for LEDCs such as Turkey rather than MEDCs such as the USA.

5 Describe the ways that people can prepare for a major earthquake.

6 Use the information in this section, together with resources from the Internet, to design an information poster about how people should prepare for and act during an earthquake. You could use a computer and desktop publishing software to produce your poster.

VOLCANOES

Volcanoes are mountains, built by successive eruptions of lava, volcanic bombs, ashflows and tephra (airborne ash and dust) which pile on top of each other. A volcano is most commonly a conical hill or mountain built around a vent that connects with reservoirs of molten rock below the surface of the earth. The term volcano also refers to the opening or vent through which the molten rock and associated gases escape.

Some volcanic eruptions are explosive and others are not. How explosive an eruption is depends on how runny or sticky the magma is. If magma is thin and runny, gases can escape easily from it. When this type of magma erupts, it flows out of the volcano. If magma is thick and sticky, gases cannot escape easily. Pressure builds up until the gases escape violently and explode. It is the type of lava that determines the shape of the volcano and how it erupts.

TASKS

1 What is a volcano?

2 Why are most volcanoes close to plate boundaries?

3 Why is the eruption of each volcano different?

4 Why are some volcanic eruptions more destructive than others?

5 You can find out more about the features of volcanoes from CD-ROMs such as *Encarta Encyclopedia*, *Physical World* and *Geodome*; and the Internet site Volcano World http://volcano.und.nodak.edu/

MOUNT ST HELENS

Cause of the eruption

Mount St Helens is a peak in the Rocky Mountain fold mountain range in North America. The mountain range was formed approximately 70 million years ago at a convergent plate boundary A minor plate, the Juan de Fuca Plate, which is oceanic crust, is moving eastwards, colliding with the North American Plate, which is continental crust The oceanic crust is forced downwards into the mantle. The increase in temperature destroys this crust, turning it into magma. Pressure increases in the mantle and the magma rises to the earth's surface. This leads to volcanic eruptions. Through the centuries a series of eruptions have occurred at this plate boundary forming the Cascade Range. Mount St Helens is one of a series of volcanic peaks in this range. Geologists call Mount St Helens a composite volcano, a term for steep-sided, often symmetrical cones constructed of alternating layers of lava flows, ash and other volcanic debris. Composite volcanoes tend to erupt explosively and pose considerable danger to nearby life and property.

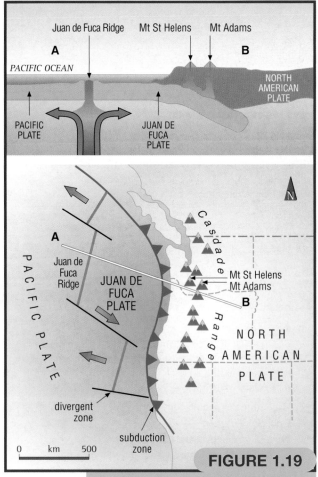

FIGURE 1.19

Case Study: Mount St Helens, Washington

The eruption

20 March, 1980, at 3:47pm Pacific Standard Time (PST), a magnitude 4.2 (Richter Scale) earthquake was the first signal of the reawakening of the volcano. A succession of smaller quakes followed over the next few days.

27 March, Mount St Helens began to spew ash and steam, marking the first significant eruption.

Early May, the northern side of Mount St Helens began to bulge by 1.5m a day. Public authorities closed the area surrounding the mountain after being informed by geologists of the volcano's past violent behaviour. This proved unpopular with some local people.

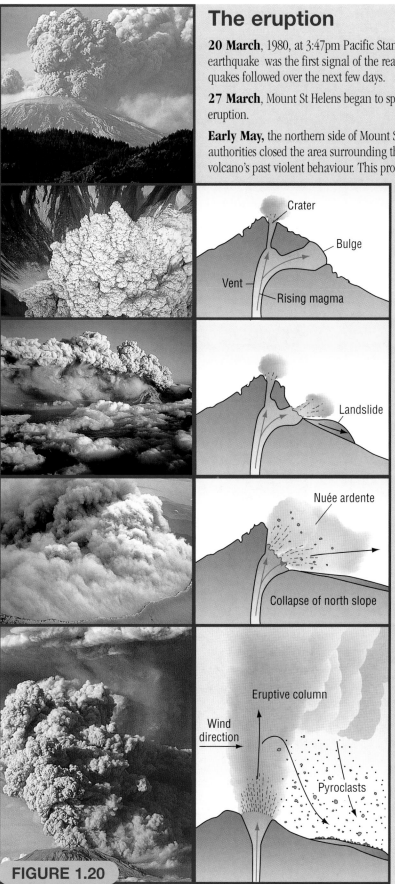

18 May 8.30am Small ash and steam eruptions were rising from the crater of the volcano.

18 May 8.32am An earthquake (magnitude 5 on the Richter Scale) caused the bulging northern slope to move forwards and downwards. This became a great landslide of soil, glacier ice, snow and rock, which raced down the mountainside to fill in Spirit Lake. This material together with the water flung out of the lake by the landslide, then moved down the Toutle valley as a mudflow. This eventually blocked the channel on the navigable Columbia River 60 km away.

18 May 8.33am The landslide exposed magma in the volcano's vent. It exploded sideways sending out a gigantic blast of hot volcanic gases, steam, dust and rock fragments, moving at speeds of over 100 kph. It felled trees up to 24 km away. Every form of life – plant and animal, within this range was destroyed. This type of blast is known as a **nuée ardente**.

Rest of the morning An eruptive column rose to an altitude of more than 20 km, depositing ash over a wide area. In the town of Yakima, 120 km away, the ash deposits were 1cm thick, the inhabitants could only go out if they wore face masks. Volcanic debris, formed from the solidifying magma, fell from the column, onto the remains of the northern slope of the mountain. These deposits are known as volcanic bombs or **pyroclasts**. Near the volcano, the swirling ash particles in the atmosphere generated lightning, which in turn started many forest fires.

Early on 19 May the eruption had stopped. By that time, the ash cloud had spread to the central United States. Two days later, fine ash was detected in several cities of the north-eastern United States. Some of the ash drifted around the globe within about 2 weeks.

FIGURE 1.20

Stages of the eruption

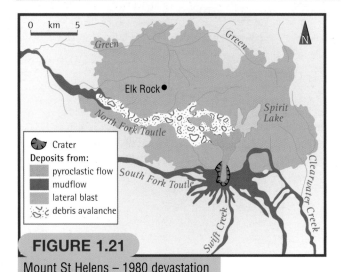

FIGURE 1.21

Mount St Helens – 1980 devastation

08:32:21 am

08:32:47 am

08:32:53 am

FIGURE 1.22

Mount St Helens as seen from Johnston Ridge, 18 May

Effects of the eruption

The 18 May 1980, eruption was the most destructive in the recorded history of the United States. Mount St Helens' eruption caused widespread destruction. The lateral blast, debris avalanche, mudflows and flooding caused extensive damage. Landscape changes caused by the 18 May eruption are clearly evident on the photographs and satellite images (page 20) The 18 May eruption resulted in scores of injuries and the loss of 61 lives. Most of Mount St Helens' victims died by

asphyxiation from inhaling hot volcanic ash, and some by burns and other injuries. More than 200 houses and cabins were destroyed, leaving many people homeless. Over 185 miles of highways and roads and 15 miles of railways were destroyed or extensively damaged as well as 27 bridges.

More than 4 billion board feet of commercial timber (enough to build about 300 000 two-bedroom homes) were damaged or destroyed, primarily by the lateral blast. At least 25% of the destroyed timber was salvaged after September 1980. Hundreds of loggers were involved in the timber-salvage operations, and, during peak summer months, more than 600 truckloads of salvaged timber were retrieved each day.

FIGURE 1.23

Mount St Helens – blowdown

Wildlife in the Mount St Helens area also suffered heavily. The Washington State Department of Game estimated that nearly 7 000 big game animals (deer, elk, and bear) perished in the area most affected by the eruption, as well as all birds and most small mammals.

FIGURE 1.24

Mount St Helens – volcanic ash

The cost of the destruction and damage caused by the 18 May eruption has been estimated at $1.1 billion.

Initially the eruption dealt a crippling blow to tourism, an important industry in Washington. Long term, however, tourism has benefited from the eruption; many people have been keen to visit the site first hand. In the early 1980s, tens of thousands of visitors flocked to the area surrounding Mount St Helens to marvel at the effects of the eruption. On 27 August, 1982, President Reagan signed into law a measure setting aside 110 000 acres around the volcano as the Mount St Helens National Volcanic Monument, the nation's first such monument. Since then, many trails, viewpoints, information stations, campgrounds, and picnic areas have been established to accommodate the increasing number of visitors each year. Many people take part in the adventure holidays offered over the Internet.

FIGURE 1.25

Mount St Helens – the morning after

Downwind of the volcano, in areas of thick ash covering, many agricultural crops, such as wheat, apples, potatoes and alfalfa, were destroyed. Many crops survived, however, in areas blanketed by only a thin covering of ash. In the long term, the ash may provide beneficial chemical nutrients to the soils of eastern Washington State.

The ash fall caused problems for transportation, because visibility was greatly reduced. Many highways and roads were closed to traffic, some only for a few hours, but others for weeks. Interstate 90 from Seattle to Spokane, Washington, was closed for a week. Air transportation was disrupted for up to two weeks as several airports in eastern Washington shut down due to ash covering and poor visibility. Over a thousand commercial flights were cancelled following airport closures.

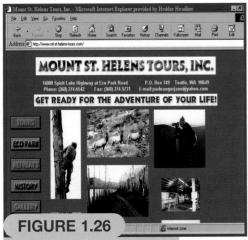

FIGURE 1.26

Mount St Helens Tours website

TASKS

1 With the aid of a diagram, explain the causes of the 1980 Mount St Helens eruption.

2 Describe the events leading up to the eruption.

3 Explain what finally triggered the eruption.

4 List the immediate and long-term effects of the eruption.

5 The following Internet sites contain an extensive collection of resources about the 1980 eruption at Mount St Helens:

Cascades Volcano Observatory
Volcano World
http://vulcan.wr.usgs.gov/home.html
http://volcano.und.nodak.edu/

If you visit these websites you can find out much more about the eruption. You could download resources from the websites into desktop publishing software to produce your own newspaper front page about the disaster.

6 The screen dump, Figure 1.26 is for an adventure holiday company in the area. Visit the website: www.mt-st-helens-tours.com/ and explain how the volcanic eruption has provided long-term advantages for the region.

VOLCANOES AND SATELLITE IMAGERY

 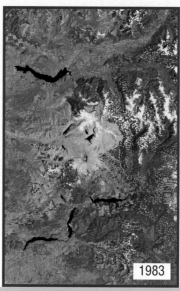 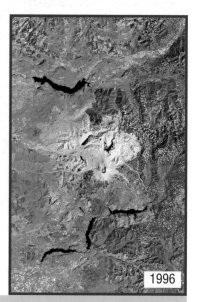

FIGURE 1.27 | 1973 | 1983 | 1996 |

Ash, mudslides, and mud-laden rivers show as greyish blue.
Water looks black.
Ice and snow are white.
Gradual vegetation regrowth, as light red and pink, in the devastated area.

The three satellite images of Mount St Helens in 1973, 1983 and 1996 were taken from an internet site in the USA:USGS Earthshots Satellite Images of Environmental Change
http://edcwww.cr.usgs.gov/earthshots/fast/tableofcontentstext

Landsat images show how much energy from the sun (electromagnetic radiation) was being reflected or emitted off the earth's surface when the image was taken. Clear water reflects little radiation, so it looks black. Pavement and bare ground reflect a lot of radiation, so they look bright. Urban areas usually look light blue-grey. Vegetation absorbs visible light but reflects infrared, so it looks red.

These satellite images show the area around Mount St Helens, in south-western Washington, before and after its eruption of 18 May 1980.

Earthshots is a collection of Landsat images and text, designed to show environmental changes and to introduce remote sensing. Images from other satellites, maps and photographs are also included. Earthshots comes from the US Geological Survey's EROS Data Center, the world's largest archive of earth science data.

The home page provides access to a collection of invaluable pages to develop the user's understanding about how to use remote sensing.

TASKS

1 Compare the 1973 and 1983 images and describe how the eruption of Mount St Helens changed the area.

2 Compare the 1983 and 1996 images and describe the changes that have occurred since the eruption.

3 How can you tell that the 1996 image was taken during the summer?

4 Using evidence from the satellite images, why do you think there is an absence of settlements in the region?

5 Use the Internet to access the Earthshots website. Find the case study on Mount St Helens, copy and paste the satellite images into desktop publishing software, and label the changes using the software tools.

You can use this Internet site when investigating other topics in this GCSE course, for example deforestation of tropical rainforests (page 186); urbanisation (page 82).

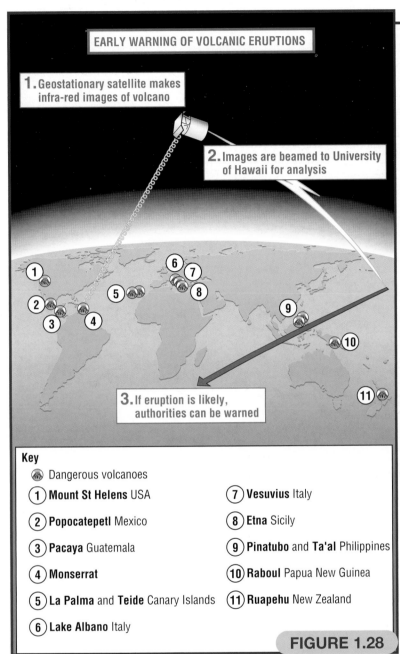

EARLY WARNING OF VOLCANIC ERUPTIONS

1. Geostationary satellite makes infra-red images of volcano

2. Images are beamed to University of Hawaii for analysis

3. If eruption is likely, authorities can be warned

Key

🌋 Dangerous volcanoes

1 **Mount St Helens** USA

2 **Popocatepetl** Mexico

3 **Pacaya** Guatemala

4 **Monserrat**

5 **La Palma** and **Teide** Canary Islands

6 **Lake Albano** Italy

7 **Vesuvius** Italy

8 **Etna** Sicily

9 **Pinatubo** and **Ta'al** Philippines

10 **Raboul** Papua New Guinea

11 **Ruapehu** New Zealand

FIGURE 1.28

Satellites spot volcano ready to erupt

A volcanic eruption has been successfully predicted with the use of satellites. Researchers disclosed yesterday that they had detected the impending eruption of Pacaya, in Guatemala, a week before it began.

The team, using satellites with infra-red detectors, picked up a heat signal on May 13 indicating that hot magma was bubbling towards the surface. The volcano erupted on May 20 sending an ash cloud over Guatemala City and the airport 13 miles away.

Luke Flynn, a vulcanologist at the University of Hawaii, said yesterday: "We've never had a way to remotely monitor volcanoes for impending eruptions before."

Andrew Harris, at the Open University in Milton Keynes, who was a member of the team, said: "We saw it coming from space. To date this has not happened before."

The breakthrough, by British and American scientists, may lead to the establishment of a worldwide automatic forecasting system for the 600 active volcanoes and many others considered to be potentially active.

The team also detected the eruption of a volcano in the remote Galapagos islands three hours before it began on September 15. The early warning gave experts on the ground time to move wildlife. The signals were picked up from the satellites by the University of Hawaii.

The team also spotted the eruption of Popocatepetl, near Mexico City, from space. The satellites detected a moderate eruption on the morning of November 24 this year. Local ground-based teams recorded the same event and sounded the alarm one minute earlier. But many parts of the world where volcanoes could burst into life are too treacherous to have trained staff in place. Dr Harris said: "Some places are just too poor and have too many volcanoes."

The satellite system, even if it spots an eruption only as it occurs, may give emergency services vital hours or days to get people cleared from an impending lava flow.

The breakthrough has been made possible by the recent launch of two geostationary satellites owned by the US National Oceanic and Atmospheric Administration. The craft can provide images of a given volcano or area of land every 15 minutes. Recent computer developments mean that the images can be rapidly analysed for hot spots. The researchers are posting the results on the University of Hawaii's Hot Spot Image Internet site.

Dr Harris said that about seven satellites, able to see heat in the right waveband, would be needed to create a global volcano early-warning system. "At the moment we are only really covering the Americas and the Caribbean."

About 360 million people live on or near dangerous volcanoes, from Elna in Sicily to Mount St Helens in Washington State. In recent years volcanoes have killed about 25,000 people.

Aircraft are also at risk. In 1982 a British Airways Boeing 747 nearly fell from the sky over Java after volcanic dust got into the jet engines.

The quest for an early warning system for volcanic eruptions has been given more urgency after indications that rising sea levels because of global warming may make volcanoes more active.

12,000 years ago, when sea levels rose by 38ft in two centuries, volcanic activity surged. Rising sea levels trigger landslips and weaken the rock, releasing pressure.

Some researchers have tried to develop early warning systems that pick up microquakes inside a grumbling volcano. Others have been developing systems that detect land movement and bulges on the earth's surface in advance of an eruption.

FIGURE 1.29

Nick Nuttall, *The Times*, 8 December 1998

TASKS

1 Which volcanic eruptions were successfully detected with the use of satellites?

2 How can satellites predict volcanic eruptions?

3 Why is it important that people obtain early warning of a volcanic eruption?

4 What other early warning systems for volcanic eruptions are being developed?

WHY DO PEOPLE LIVE NEAR VOLCANOES?

It has been estimated that 500 million people now live at risk from volcanic hazards. In the past 500 years, over 200 000 people have lost their lives due to volcanic eruptions. An average of 845 people died each year between 1900 and 1986 from volcanic hazards. The number of deaths for these years is far greater than the number of deaths for previous centuries. The reason behind this increase is not due to an increase in volcanic activity, but due, instead, to an increase in the amount of people populating the sides of active volcanoes and valley areas near those volcanoes. There are a variety of advantages of living in areas of volcanic activity, which increasingly outweigh the risk of coping with the effects of the remote possibility of an eruption. As vulcanologists begin to develop more sophisticated and accurate methods of predicting imminent eruptions, the risk of living in volcanic areas is further reduced.

The Plus Side of Volcanoes

Fertile Soils

As you have already seen, volcanoes can cause much damage and destruction, but in the long term they can also benefit people. Over thousands of years, the physical breakdown and chemical weathering of volcanic rocks have formed some of the most fertile soils on earth. In tropical regions, such as the island of Hawaii, the formation of fertile soil and growth of lush vegetation following an eruption can be as fast as a few hundred years. The slopes of Vesuvius in Southern Italy have weathered to produce fertile soils which are used to grow a wide range of products including olives, fruit, vines and nuts. One in five Sicilians lives on the slopes of Mount Etna to farm on the fertile soils even though the volcano erupts on average once every ten years. Some of the best rice-growing regions of Indonesia are in the shadow of active volcanoes (Figure 1.30).

Geothermal energy

Geothermal energy can be harnessed from the earth's natural heat associated with active volcanoes or geologically young inactive volcanoes still giving

FIGURE 1.30

A lush rice paddy in central Java, Indonesia; Sundoro Volcano looms in the background. The most highly prized rice-growing areas have fertile soils formed from the breakdown of young volcanic deposits.

off heat at depth in the crust. Geothermal energy can usually be generated where water stored in permeable rocks is heated by magma to form superheated steam. The steam is tapped by boreholes drilled into the permeable rocks. This steam can be used to drive turbines in geothermal power stations to produce electricity for domestic and industrial use. This method of generating electricity is used successfully in countries like New Zealand and Iceland. Geothermal heat warms more than 70% of the homes in Iceland. Steam from high-temperature geothermal water can be used to drive turbines and generate electrical power, while lower temperature water provides hot water for space-heating purposes, heat for greenhouses and industrial uses, and hot or warm springs at resort spas. The great advantages of this form of energy are that they are environmentally friendly and almost limitless.

Minerals

Lava in volcanoes can crystallise to form minerals, like gold, silver, diamonds, copper and zinc, depending on their mineral composition. Therefore, dormant or extinct volcanoes are excellent places for mining, creating job opportunities leading to the development of mining towns near volcanoes.

Tourism

People throughout the world have a fascination with volcanoes, which is successfully utilised by tourist operators worldwide. Many tourists flock to places like Hawaii and Mount Merapi to catch a glimpse of flowing lava or smoke billowing out of volcanoes. This brings in revenue for the country. Hot springs and geysers can also bring in the tourists. Many health spas use the natural hot springs and volcanic materials near volcanic areas to attract tourists, since many believe being covered in volcanic materials and soaking in hot springs is not only relaxing, but can also cure some ailments. Old Faithful, a geyser in Yellowstone National Park in the USA, which erupts at regular intervals, is a famous tourist attraction. Since the early 1980s, tens of thousands of visitors have flocked to the area surrounding Mount St Helens to marvel at the effects of the eruption. A Visitor Center was completed in December 1986 at Silver Lake, about 30 miles west of Mount St Helens; by the end of 1989, the Center had hosted more than 1.5 million visitors.

Monitoring and predicting eruptions

The main reason scientists study and monitor volcanoes is so that those living near active volcanoes can be aware of the hazards they produce. It is important that scientists communicate with local government officials and the general public about hazards produced by the volcanoes in their area, hopefully saving lives and encouraging better land use planning. The problem with volcanoes is that, though there may be similarities between volcanoes, every volcano behaves differently and has its own set of hazards. When scientists study volcanoes, they map past volcanic deposits and use satellites to look at volcanic features, ash clouds, and gas emissions. They also monitor seismic

FIGURE 1.31

The eruption of Popocatepetl in Mexico in November 1998 was predicted from satellite evidence and warnings were issued.

activity and thermal changes. They study and monitor volcanic gases and monitor the temperature, flow rate, sediment transport, and water level of streams and lakes near the volcano.

By studying volcanic deposits, scientists can produce hazard maps. These maps indicate the types of hazards that can be expected in a given area the next time a volcano erupts. Dating of these volcanic deposits helps determine how often an eruption may occur and the probability of an eruption each year. Monitoring of a volcano over long periods of time will indicate changes in the volcano before it erupts. These changes can help in predicting when an eruption may occur.

TASKS

1 Explain the reasons why people continue to live in areas near active volcanoes.

2 How can scientists minimise the risks faced by people living near volcanoes?

THE HYDROLOGICAL CYCLE

The **drainage basin** is an area of land which collects rainwater and feeds it into a network of streams. The boundary of the basin is the **watershed**.

The water balance

A drainage basin can have a water balance in the same way that a building society account has a money balance at the end of a year after credits (inputs) and debits (outputs) have taken place. It is possible to write a balanced equation for a drainage basin, using the annual precipitation, runoff and evapotranspiration totals. (**Runoff** is an estimate of how much water leaves the basin at the river's mouth and **evapotranspiration** is an estimate of how much water is returned from the basin to the atmosphere.) Water balances for two British river basins are shown below:

	precipitation	= runoff + evapotranspiration
River Ystwyth	1563mm	= 1289mm + 274mm
River Yare	739mm	= 132mm + 607mm

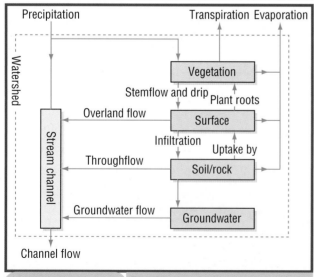

FIGURE 1.33   The drainage basin as a system

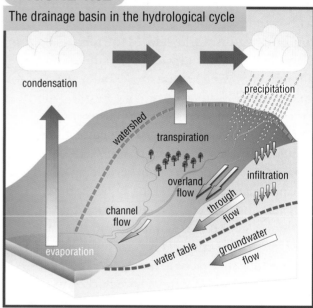

FIGURE 1.32

The drainage basin in the hydrological cycle

The storm hydrograph

A **hydrograph** is a graph of stream discharge against time; the word 'storm' refers to the rainfall event that causes the stream discharge to increase.

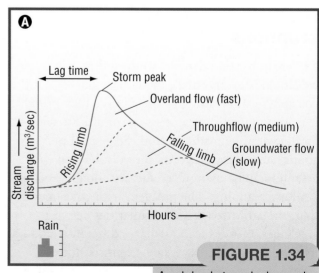

FIGURE 1.34

A subdued storm hydrograph

Analysing storm hydrographs

Hydrographs vary in the detail of their shapes. Figures 1.34 and 1.35 show two hydrographs with several differences:

- A has a longer **lag time** than B. The lag time is the period between the rainfall event and the peak stream discharge.
- B is a flashy hydrograph whereas A is subdued.
- A's response to the rain lasts longer then B's.
- B has a higher peak discharge than A.
- B's rising and falling limbs are steeper than A's.

FIGURE 1.37

The old Wye bridge, Hereford, at low flow.

FIGURE 1.36

The old Wye bridge, Hereford, at high flow.

These variations could occur between different streams or on the same stream at different times. Some of the factors that cause the variations are:

- **land use:** vegetation – especially trees – will intercept rain so that less water arrives at the ground surface, which reduces the possibility of, and rate of overland flow. It makes a subdued hydrograph more likely. The concrete and tarmac which cover much of the surface in an urban area do not allow water to soak in and this tends to encourage overland flow and a flashy storm hydrograph.

- **type of rain:** a heavy downpour is likely to mean that some of the rain will not be able to infiltrate the ground. As the rainfall rate is greater than the infiltration rate, this will create more overland flow and a flashy storm hydrograph.

- **soil moisture:** rain falling onto a saturated soil will not all be able to infiltrate and this may encourage overland flow and a flashy storm hydrograph.

B Most of the water in a flashy hydrograph has arrived at the river quickly as overland flow; most of the water in a subdued hydrograph has arrived more slowly as throughflow and groundwater flow.

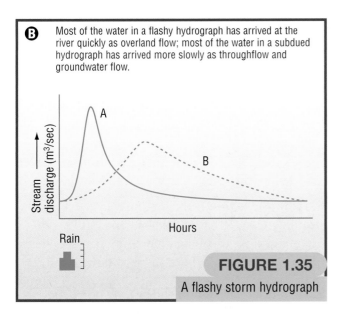

A

B

Stream discharge (m³/sec)

Hours

Rain

FIGURE 1.35

A flashy storm hydrograph

TASKS

1 Redraw Figure 1.33 using the following symbols drawn alongside each word as appropriate:

INPUT ↓ OUTPUT ↑
STORE ◿ FLOW ∿

2 Explain why the Ystwyth and Yare river basins have such different water balances. You will need to use an atlas to locate the rivers and also to find out about their climate and land use.

3 Describe and explain the characteristic shape of the storm hydrograph.

4 Arrange the following words in pairs under the correct heading: **flashy   subdued**
rural   ploughed   wooded   thunderstorm   dry soil
urban   drizzle   saturated soil

RIVER PROCESSES

From its headwaters to its mouth a river channel is a system. Energy comes from water flowing down towards sea level, although much of the running water's energy is used to overcome friction from the river's bed and banks.

A river's energy enables it to make changes.

- **Erosion** occurs when the running water wears away its bed and banks. **Vertical erosion** occurs when the river cuts downwards into its bed and **lateral erosion** occurs when the river cuts sideways into one or other of its banks. The amount of erosion increases when the energy of the running water increases – as it flows faster. Much erosion occurs in times of river flooding, when the water is flowing fastest.

FIGURE 1.38

The Holford River in its middle course

- **Transport** occurs when the river carries material (**load**) that it has already eroded. The size and number of particles that can be carried depends on the velocity of the river.

- **Deposition** occurs when the river loses energy, stops eroding and lays down load it has been carrying.

Erosion processes

A river erodes when the following processes are at work:

- **Abrasion** is the process by which solid particles carried by the flow of the river are thrown and rolled against the bed and banks with sufficient force to wear them away. Evidence of this process is provided by **potholes** – round hollows in a rocky river bed which contain a collection of rounded stones that have been swirled around by an eddy current (whirlpool) and have effectively drilled the hollow into the bed.

- **Attrition** is the wearing down of particles which the river transports; as the particles strike the bed, banks or other particles their sharp corners are knocked off and they gradually become rounder and smaller.

- **Solution** is a chemical process in which the weak acids contained in the river water dissolve the minerals that make up the rock and soil of the river's bed and banks. Some rocks – such as limestone – are relatively easy to dissolve and some rivers are relatively acid, such as those which are fed by the water which drains from peat bogs.

- **Hydraulic action** is the hammering effect which turbulent water has on the bed and banks – this hammering is capable of loosening particles that are eventually pulled out into the flow of water.

Erosion occurs fastest on the outside of a bend, where the water is flowing fast (see Figure 1.38).

TRANSPORT PROCESSES

A river transports material in four ways:

▶ **Dissolved load** is the material which cannot be seen but is dissolved in the river water. The amount of material that is transported in this way is much smaller than the other three ways.

▶ **Suspended load** is the material which is light enough to be carried by the river's current – it is the suspended load that makes swollen rivers look 'muddy'.

▶ **Bedload** is material which is too heavy to be carried by the river's current and is rolling along the river bed, dragged by the force of the moving water.

▶ **Saltation load** is material which moves by bouncing along the river bed in short 'hops', so its movement process is somewhere between bedload and suspended load.

The last three methods of transport depend on the speed of the river's flow: a particle may be lying still on the river bed but if the river's speed increases (as it will in a rising storm hydrograph), the particle may start rolling [bedload] before bouncing [saltation load] and then finally taking off into the flow [suspended load].

RIVER DEPOSITION

If the speed of the moving water decreases it will lose energy and will no longer be able to carry out so much erosion and transport. As the river's speed decreases particles will be deposited (laid down) in reverse order of size, i.e. larger particles will be deposited first. There are two main causes of a decrease in river speed:

▶ the river's volume has passed its storm peak and is on the falling limb of its hydrograph;

▶ there has been an increase in friction due to the flow entering shallower water or a restriction such as vegetation.

In both cases the river will start to deposit material which it was carrying and this may build up in large quantities; this often happens on the inside of a bend (see Figure 1.38).

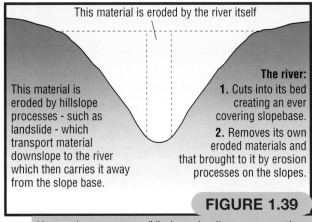

FIGURE 1.39

How a river creates a 'V'-shaped valley cross-section

This material is moved by hillslope processes, such as landslides, which transport material downslope to the river which then carries it away from the base of the slope.

The river is therefore doing two things:

▶ cutting into its bed to create an ever-lowering slope base;

▶ removing its own eroded material and that which has been brought to it by erosion processes on the slopes beside it.

TASKS

1 What do you think is the relationship between:
   a the roundness of the particles in the bedload *and* the distance downstream;
   b the size of the particles in the bedload *and* the distance downstream;
   c the velocity (speed of flow) *and* distance downstream;
   d the river velocity *and* the area of a cross-section of the river's channel;
   e river load *and* river speed.

2 a Draw the river bend shown in Figure 1.38. Add an arrow to show the fastest water flow through the bend.
   b Annotate your diagram to show the following features:
   line of fastest water flow;
   steep bank;
   gently sloping bank;
   erosion causing undercutting;
   material deposited in slow-moving water.

3 Some rivers have regular high-energy floods followed by long periods of low flow. During these long periods of low flow the river has to pick its way between islands of material which it hasn't the energy to carry. This is called a **braided channel**.
   a Draw a plan to show what a braided channel looks like;
   b Suggest two reasons why a river's flow may vary.

RIVER LANDSCAPE FEATURES

The river landscape changes as the water flows from its upper course, through its middle and lower courses and – eventually – into the sea. Features of the landscape are:

Upper course
waterfalls
rapids
large, angular bedload
potholes in bare rock bed
steep gradient
'V'-shaped valley
interlocking spurs

Middle course
small meanders
narrow floodplain
moderate gradient

Lower course
more sinuous meanders
wide floodplain
gentle gradient
levées, ox bow lakes

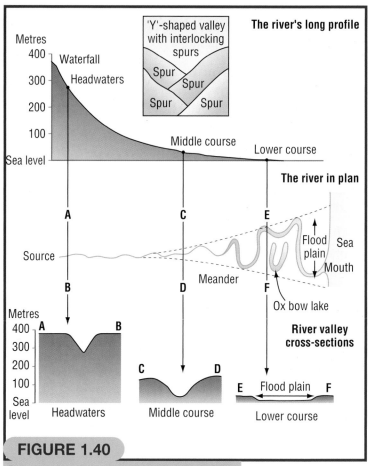

FIGURE 1.40

Downstream changes along a river's course

UPPER COURSE FEATURES

In its headwaters, the river's energy is concentrated on cutting down vertically into its bed, which means that the bed is often made of eroded bare rock. A major way in which the river erodes its bed is by wearing away potholes, as described on page 26.

The river's course is fairly straight in its upper course because there is little lateral (sideways) erosion to make it wander from side to side. Looking along the headwaters the stream appears to wind gently at the bottom of a 'V'-shaped valley between interlocking spurs.

Waterfalls are sometimes found in the upper course of a river where it crosses a hard layer of rock lying on top of softer rock. Figure 1.41 shows how the soft rock is worn away more than the hard rock which eventually collapses because it is no longer supported. The waterfall gradually retreats upstream from its original position, leaving behind a narrow, steep-sided gorge.

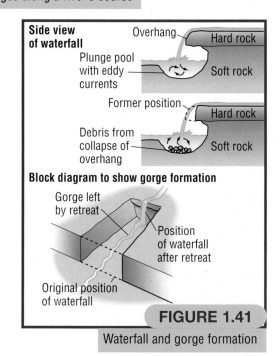

FIGURE 1.41

Waterfall and gorge formation

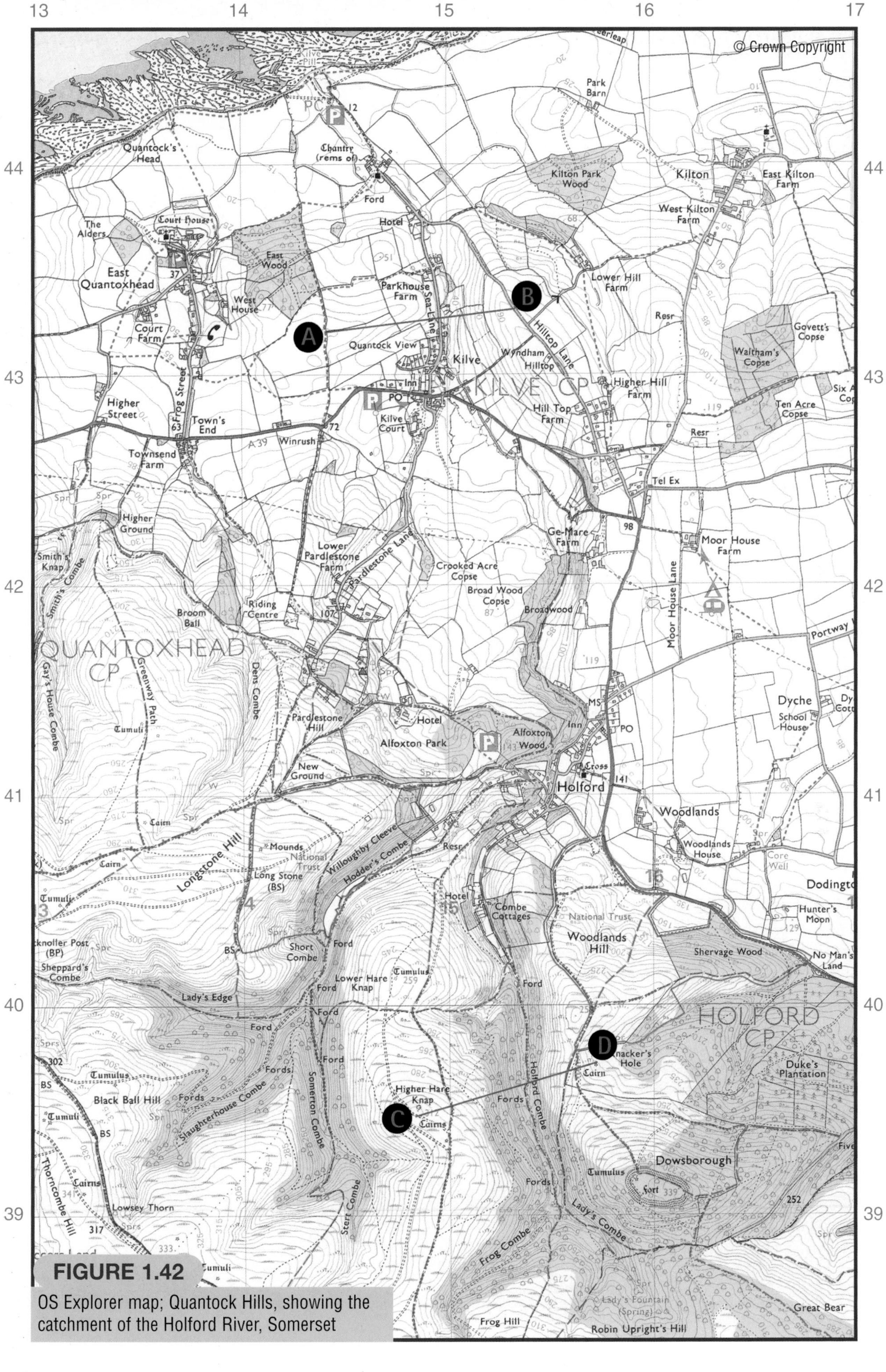

FIGURE 1.42

OS Explorer map; Quantock Hills, showing the catchment of the Holford River, Somerset

© Crown Copyright

LOWER COURSE FEATURES

In its lower course the river's energy is concentrated on cutting sideways into its banks. This means that it creates a flat floodplain across which the river winds in loops known as meanders (see Figure 1.43).

Running water flows fastest on the outside of a bend, as Figure 1.46 shows; the lines on this diagram are isovels – lines joining points of equal velocity, measured in metres per second. The fast-flowing water on the outside of the bend causes erosion while the slower flow on the inside of the bend encourages deposition.

Occasionally, a great flood may cut through the narrow neck of a meander and when the flood goes down the river may follow a new course (see Figure 1.45). The old course of the river is left as a curved pool known as an **ox bow lake**.

Every time the river floods above its banks there will be a sudden drop in water velocity as the fast-moving channel flow meets the frictional drag of the floodplain. The tops of both river banks are favoured places for deposition to occur, forming low ridges known as **levées**).

The course taken by the fast-moving water through a series of bends creates erosion of their downstream, outside banks and deposition against their upstream, inside banks. This results in meanders gradually migrating downstream, slowly widening the floodplain as they go.

FIGURE 1.43

The River Swale in its lower course

Formation of an ox bow lake

1

Floodplain

Neck of meander

X
Y

Meander

2

Ox bow

Formation of levées

Floodplain — Floodplain

Normal flow

Water velocity reduced due to friction

Flood conditions

Reduced velocity encourages deposition

Levées formed

Flood subsides

Process repeats at every flood and levées grow in height

FIGURE 1.44 Floodplain features

TASKS

1 Make a large, simple line drawing of Figure 1.43 and annotate it to show the following features:
a meander
the ox bow lake
where another ox bow lake may form.

2 Using the 1:25 000 Ordnance Survey map on page 29, draw cross-sections of the valley of the River Horner in its upper and lower courses, along the lines marked on the map. The vertical scale for your cross-section should be 1cm is equivalent to 100 m (1:10 000); this is a vertical exaggeration of x 2.5.

3 Using the 1:25 000 Ordnance Survey map on page 29 draw a long profile of the Holford river from its upper course at Holford Combe (154400) to its mouth at Kilve Pill (143444). You should use the same technique as you would to draw a cross-section but you will have to keep turning the straight edge of paper to accommodate the river's winding course. The vertical scale for your long profile should be 1cm is equivalent to 100 m (1:10 000); this is also a vertical exaggeration of x 2.5.

4 Draw an annotated diagram to show how meanders migrate downstream.

5 You could investigate river features further using CD-ROMs such as *Geodome*, *Physical World* and the *Environment Agency Riverside Explorer* CD-ROM.

If you live near a river use a digital camera to take pictures of different river features. Download the images and save them into a folder either on your computer or a floppy disc.

Insert your digital images into a desktop publishing program, use the software's tools to label the characteristic features of the river.

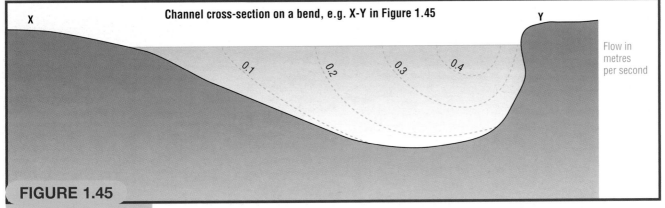

FIGURE 1.45

Diagram of river velocity

WHY RIVERS FLOOD

A river flood is when a river spills out over its banks and flows onto areas which are not usually covered by water. The main causes of river floods are:

▶ heavy rainfall

▶ rapidly melting snow

▶ soil saturation

▶ deforestation (removing trees)

▶ ploughing

▶ dam bursts

▶ urbanisation (extending built-up areas).

FLOODING IN THE NETHERLANDS, 1995

Natural causes

In early 1995, there was heavy rain over much of Europe. In some parts, it was almost continuous from November 1994 to February 1995, and Switzerland received three times its January average. Snow melted early and quickly in the Alps. The ground was saturated because of the heavy rain so further rain was rapidly transferred to rivers as overland flow.

Some people feel that the floods are evidence of the effects of global warming. In the last 100 years:

▶ average temperatures have risen by about 1°C in southern Germany.

▶ winter precipitation in the Rhine catchment has increased by 40%.

FIGURE 1.46

Flooding in the Netherlands (note the protective dyke around the farmhouse)

Human causes

▸ Pressure for use as farmland or building means that the Rhine has lost much of its riverside marsh and floodplain which used to hold back floodwater.

▸ Improved flood protection measures upstream, such as higher embankments, mean that floodwater moves downstream more quickly than it used to.

▸ Improved navigation for shipping has involved straightening the river (its journey of 1 320 km from source to sea has been shortened by 50 km) – so water moves downstream more quickly.

▸ The upper Rhine is used for generating hydro-electric power; ten power stations are bypassed by an efficient new channel parallel to the old river, designed to remove water which is surplus to the generators' requirements.

▸ A flood surge used to be spread over five days but now occurs over just two days: the same volume of water is moving further in a shorter time causing a dramatic rise in the river's level.

▸ The shift from pastoral to arable land use in rural areas has led to the removal of hedgerows and meadows and their replacement with ploughed fields with a reduced capacity for interception and infiltration; a greater percentage of the rain falling on the Rhine catchment enters the river.

▸ Urbanisation in the Rhine catchment has led to a threefold increase in its built-up area; the concrete and tarmac send more water to the river than the fields which they replaced.

The Rhine basin covers 220 000 square km and is home to 40 million people. In its lower course the Rhine flows through the Netherlands in many distributaries, the main one of which is called the Waal.

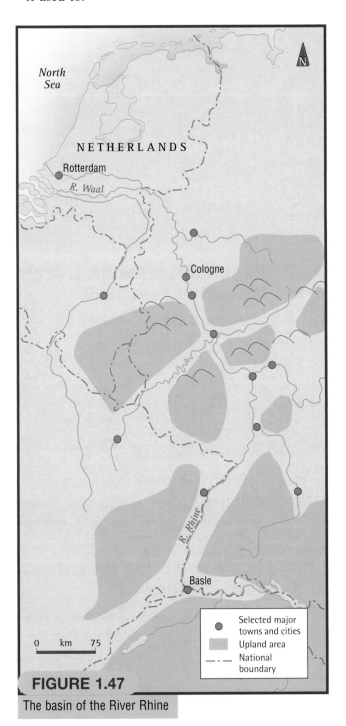

FIGURE 1.47

The basin of the River Rhine

Legend:
● Selected major towns and cities
▨ Upland area
– · – National boundary

0 km 75

TASKS

1 Using pages 24 and 25 to help you, explain how each of the following points may cause a river to flood; you should use the correct technical terms wherever appropriate in your explanations:

   soil saturation
   deforestation
   ploughing
   urbanisation.

2 Look at Figure 1.46. How has the flood risk been reduced?

3 Trace the map in Figure 1.47 and use an atlas to name and label:

   major towns and cities
   upland areas
   tributaries of the River Rhine
   the countries shown on the map.

The Netherlands flood, 1995

On 31 January 1995, the Rhine was bursting its banks at the point where it enters the Netherlands. Much of the land consists of polders – low-lying areas enclosed by protective embankments called dykes; the water table is kept below the surface of the polders by continuous pumping.

▶ Many of the polders were flooded.

▶ Four people were killed.

▶ Some roads became impassable.

▶ Many of the protective dykes are made of sand and clay which became saturated because of prolonged high river levels; this made them more likely to collapse so that emergency reinforcement work had to be carried out.

▶ 250 000 people were evacuated. Police and soldiers guarded the empty houses against looters.

▶ Many homes were flooded.

▶ Greenhouses were flooded and stocks of flowers, fruit and vegetables were lost.

▶ One million cattle were evacuated which led to some of them being infected with foot rot and reduced milk yields because of the disturbance.

▶ Waterways were closed to ships for two weeks. Many oil and dry bulk barges were stranded.

▶ Flood damage cost millions of pounds.

▶ A full-scale disaster was avoided because the authorities were well-prepared and emergency procedures were effective.

FIGURE 1.48

Temporary flood protection measures in use

Figure 1.49 Rainfall upstream at Cologne, Germany and discharge along the River Rhine from mid-January to mid-February, 1995

Date	Two-day total rainfall at Cologne (mm)	Discharge along the River Rhine (m³/second)
15.1.95	0	2 200
17.1.95	0	3 500
19.1.95	5	2 500
21.1.95	0	2 200
23.1.95	18	3 000
25.1.95	19	3 000
27.1.95	22	6 800
29.1.95	16	6 500
31.1.95	0	9 500
2.2.95	0	11 500
4.2.95	3	11 000
6.2.95	1	8 000
8.2.95	9	6 000
10.2.95	7	4 200
12.2.95	0	4 000

Short-term flood relief measures in the Netherlands in 1995

▶ sandbags and temporary barriers across doors and windows

▶ evacuation of people and livestock

▶ removing carpets and furniture to higher floors

▶ constructing temporary dykes

▶ clear underground car parks

▶ clear subways and underpasses

▶ close roads at risk of flooding

▶ install portable pumps

▶ seal door and window frames with putty or foam.

Long-term flood relief protection in the Netherlands

▶ Recognise the need for international co-operation in managing the Rhine basin to control flooding in the river's valley.

▶ Encourage afforestation (tree planting) in the Rhine drainage basin in a bid to increase the amount of rain which is intercepted.

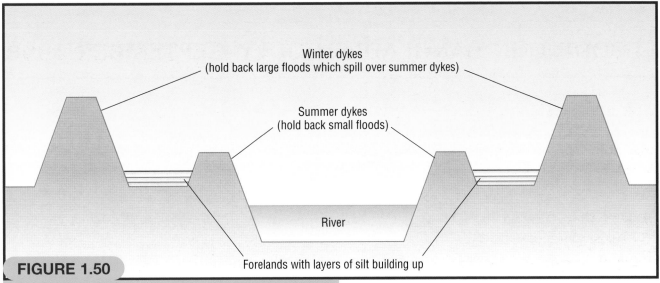

FIGURE 1.50

Cross-section of part of the Rhine floodplain in the Netherlands

Labels within figure:
- Winter dykes (hold back large floods which spill over summer dykes)
- Summer dykes (hold back small floods)
- River
- Forelands with layers of silt building up

- Encourage land uses in the Rhine basin which increase absorption of rainwater, e.g. contour ploughing and increasing the area of parks and gardens in urban areas.

- £5 billion was been spent on a system of protective dykes after floods drowned 1 800 people in 1953.

- A further £1 billion worth of flood protection was planned after the 1995 floods.

- An early warning system should improve public confidence and response but only if its predictions are accurate and the warnings are communicated effectively.

- Line earth dykes with stone blocks to reduce erosion of the dykes by abrasion during times of flood.

- Reinforce earth dykes with steel piling.

- Build flood retention basins: these are areas of riverside land which are enclosed with dykes and which have floodwater directed onto them in an attempt to reduce the river's level – when the flood subsides the water from the basin is slowly released into the channel.

- Allow the river to flow back through former marshland areas that had been sealed off for navigation purposes; although this absorbs more floodwater and creates a haven for wildlife it may slow the speed of water movement in the main channel and this may encourage the deposition of silt (fine sediment) on the river bed.

- Limit residential development in areas which are likely to flood.

- Remove the silt from the forelands (see Figure 19) which would otherwise slowly lose their capacity to hold floodwater as one flood after another deposits silt (the silt which is removed may be used to build new dykes or to make bricks).

- Encourage individual households to take steps to reduce flood damage, e.g. tiled floors downstairs with removable items of furniture and things like kitchen cupboards fixed to the wall, not resting on the floor.

TASKS

4 a Use the data in Figure 1.49 to draw a line graph of the discharge of the River Rhine and a bar graph of the two-day rainfall totals; the two graphs should share the same horizontal time axis. If you have access to a computer, you could produce the graphs using spreadsheet software, inserting the charts into a word processor to complete tasks **b** and **c** below.

   b Describe the patterns of rainfall and discharge and attempt to explain the relationships between them.

   c Suggest what probably happened to the discharge of the River Rhine in the second half of February.

5 Draw up a charter of good practice for international co-operation on flooding of the River Rhine; your charter should be in the form of a list of land-use practices which should help to reduce flooding.

The website of the International Commission for the Protection of the Rhine (ICPR) http://www.iksr.org/icpr/welcome.html will help you.

6 Imagine that your home is likely to flood occasionally to a ground floor depth of 10 cm. Describe the changes which your household could make so that the damage to your home was kept to as little as possible.

FLOODING IN BANGLADESH, JULY–SEPTEMBER 1998

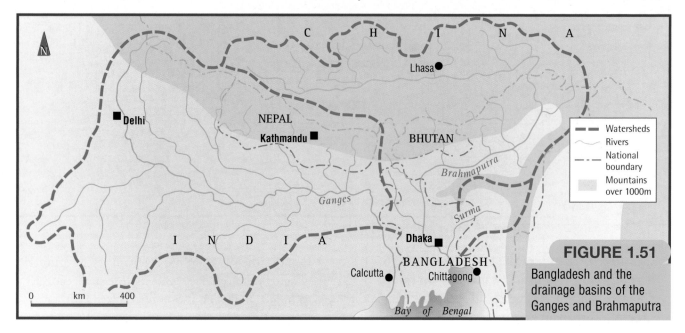

FIGURE 1.51

Bangladesh and the drainage basins of the Ganges and Brahmaputra

Natural causes

▶ Melting snow in the Himalayas was adding water to the Ganges and Brahmaputra rivers which flow through Bangladesh. The monsoon (seasonal) rains in the region were exceptionally heavy from July onwards.

▶ More than half of Bangladesh is 5m – or less – above sea level.

▶ The country lies across a huge river delta (a low-lying flat area formed from sediment which has been deposited by a river as it enters the sea); the delta contains many distributaries of the rivers Ganges and Brahmaputra which are fed by large, rainy drainage basins.

▶ Some flooding is normal. It benefits the economy of the area by providing fresh nutrients in the deposited river silt and by increasing the soil depth.

Human causes

▶ Some 92% of the area of the drainage basins which feed water into Bangladesh is in other countries (see Figure 1.52).

▶ Flooding in Bangladesh is often blamed on deforestation in the Himalayan foothills so more rainwater enters the Ganges and Brahmaputra because there is less interception by tree leaves.

▶ Some experts believe that the scale and effect of deforestation has been exaggerated.

▶ The Ganges has been diverted for irrigation purposes (bringing water to fields of crops). This has the effect of removing silt from the river's load so that when it floods further downstream in Bangladesh it no longer builds up the floodplain in that country by depositing silt.

▶ Bangladesh's rapidly expanding population needs fresh water. It comes from over 100 000 new wells. This has lowered the water table and caused the land to subside (sink) at 2.5 cm per year.

▶ Bangladesh is heavily in debt and money is not available for flood protection measures.

▶ Private investors from overseas want to see quick profits, not long-term investment in flood protection.

▶ Bangladesh needs to export more. Money gets diverted into supporting exports rather than being spent on building and maintaining flood defences.

▶ Many of the existing flood defences do not work.

▶ Corruption is taking money away from flood protection schemes.

The effects of flooding in Bangladesh

FIGURE 1.52
Muslim women wade through floodwaters in Dhaka

The immediate death toll was relatively low but the damage was devastating:

130 million cattle killed

1040 people dead in floods

Two-thirds of the country flooded

23.5 million people homeless

Worst floods this century

2½ million farmers affected

More than 1000 schools damaged

6500 bridges damaged

11 000 km of roads damaged

Some areas underwater now for two months

Dhaka residents queueing for emergency meals.

After the flood

Health

A Dhaka doctor said: 'Tackling the health problem is the biggest post-flood worry'. Water supplies were contaminated for a quarter of a million people because of polluted wells, flooded latrines and floating bodies of people and cattle; young children and the elderly were most at risk. The problems increased because of the build up of waste which could not be disposed of while the floodwaters remained. A quarter of a million people were affected by diarrhoea. Health problems were made worse by thousands of people having to live in crowded flood shelters.

Relief

Relief efforts were difficult because of the damage to communication links: the main port of Chittagong was closed for weeks, and roads and railways were cut.

Economy

Over 400 clothing factories closed. Production was down 20% in Bangladesh's important export industries, shrimp and garments. Shipping was impossible through Chittagong, the main port. Electricity supplies were also disrupted.

TASKS

1   Compare the effects of flooding in Bangladesh with those in the Netherlands (see page 34–35). Use the headings below to organise a series of short points condensed from the two pages:

Comparing causes of flooding in two countries		
Particular to Bangladesh (LEDC)	Common to both countries	Particular to the Netherlands (MEDC)

2   Explain why the effects of flooding in Bangladesh in 1998 were so much worse than in the Netherlands in 1995. Use these headings:

Differences in the physical geography of the two countries;

Differences in the wealth of the two countries.

Short-term flood-relief measures in Bangladesh in 1998

▶ 350 000 tonnes of cereal were bought by the government to feed people.

▶ The UK government gave £21 million.

▶ More than a million tonnes of international food aid was given.

▶ The World Health Organisation appealed for £5.2 million to buy water purification tablets for up to 35 million people.

▶ The Bangladesh government gave free seeds to farmers.

▶ Engineers and volunteers worked long hours to repair the flood damage. Embankments were repaired and over half of the contaminated wells were usable by mid-September.

Long-term flood protection measures in Bangladesh

▶ 7 500 km of flood embankments have been built since 1947. After a flood in 1988, a Flood Action Plan was drawn up and more embankments were built. They did not make much difference in 1998 however.

▶ Plans to cope with future floods have been made to give warnings and organise rescue and relief services.

▶ After the 1998 floods, a 50 km embankment around the city of Dhaka was proposed, but no money was available to build it.

▶ Shelters were constructed (see Figure 1.54).

▶ Dams were proposed upstream from Bangladesh to hold back some of the peak flow but the cost would be enormous.

▶ Flood retention basins similar to those on the Rhine (see page 35) were proposed but many would be needed.

▶ It was suggested that the water table in the Himalayas could be lowered by crop irrigation during the eight-month dry monsoon, providing extra capacity to absorb floodwater.

TASKS

More detailed information about this flood disaster can be found at a website constructed by the Bangladeshi Government. The site includes maps, photographs and statistics about the flood.
http://www.bangladeshonline.com/gobflood98/

3 a Use the information about the 1998 flood disaster found on the website to produce a newspaper report using desktop publishing software. Include articles outlining the causes and effects of the disaster. Illustrate your article with photographs and maps from the website.

  b You could investigate any recent flooding in Bangladesh, or attempts to control further floods by searching the online Bangladesh newspaper, The Daily Star.
http://www.dailystarnews.com

FIGURE 1.54

A flood shelter

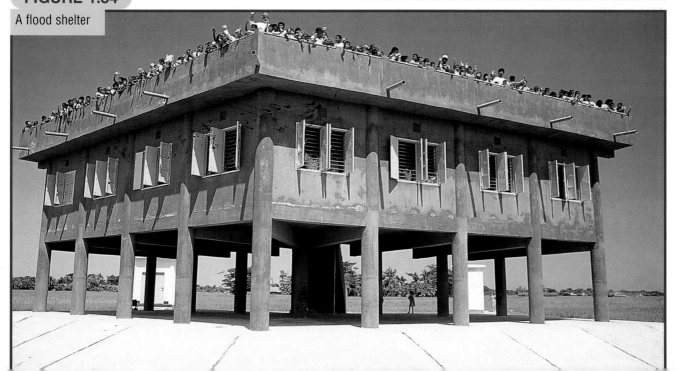

WAVE POWER

Coastal areas are in a state of constant change. The relentless impact of waves and daily tides wear down coastal features, and cause the movement of sediment and development of depositional features. Sometimes these changes are hardly noticeable but extreme weather conditions can cause rapid changes to a coastline.

Waves

Waves are formed by wind blowing over the surface of the sea. This exerts a drag or friction on the water, creating a swell in the water. The energy of the wind causes the water particles to rotate as the wind passes over. The wave moves forward, but the water particles return to their original position. The size and strength of the wave depends on three factors:

▶ The strength of the wind. The stronger the wind, the bigger the wave.

▶ How long the wind has been blowing. A prolonged wind means more energy is transferred to the wave.

▶ How far the wave has travelled. This is called the **fetch**. The longer the fetch, the larger the wave is likely to be (Figure 1.55).

When the wave reaches shallow water, the rotating water particles cannot complete their circle. Friction between the sea bed and the wave slows

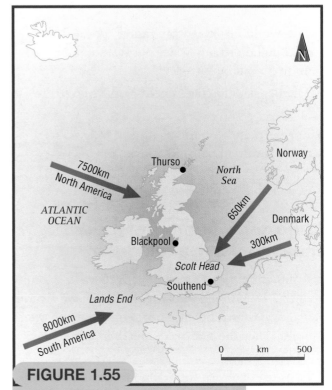

FIGURE 1.55

Fetch for different parts of the British Isles

the wave. The water begins to pile up and the tops of the waves begin to break as the wave becomes unstable and topples forward (Figure 1.56).

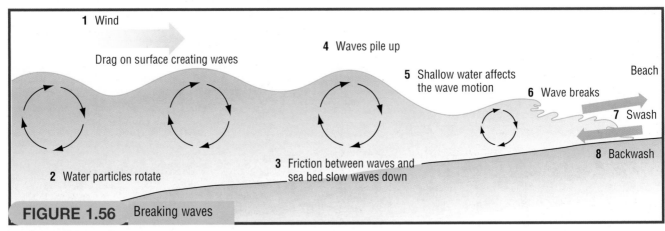

FIGURE 1.56   Breaking waves

FIGURE 1.57

Wave refraction at St Ives, Cornwall

As the sea becomes shallower, the waves alter direction to reflect the shape of the coastline. This process is called **wave refraction**. When a wave approaches a coastline with bays and headlands, the headlands slow the wave. The wave, though, continues to roll at a faster rate in the bay. Wave refraction concentrates the wave energy on the headland (Figure 1.58).

As the wave reaches the coast, it breaks and washes up the **beach**. This is called the swash. There is a slight seaward incline so the water flows back down the beach. This is called the backwash. The rate at which waves arrive at the coast determines whether the main process will be erosion or deposition. Waves which arrive at a rate of more than eight per minute, usually caused by a strong winds over a long fetch, will cause erosion. Waves arriving at a rate less than eight have less energy and will tend to deposit sediment and build up a coastline.

TASKS

1 List three factors that affect the size of a wave.

2 Describe how a wave is formed.

3 Use Figure 1.57 to explain what happens to a wave approaching the coast.

4 Draw a diagram to explain the difference between the swash and backwash.

5 How does the shape of the coastline affect incoming waves?

6 Why should the fetch of a wave and prolonged strong winds cause a wave to be erosive rather than depositional?

7 Using a globe or atlas, measure the maximum fetch for these coastal areas in kilometres:

Penzance

Aberystwyth

Cromer

Thurso

Dover

Sligo

Galway

Which side of the British Isles is likely to experience the most powerful waves?

WEARING THE COAST AWAY

Waves crashing into the coast exert considerable force. Research estimates pressures upwards of 30 tonnes per square metre can occur. This figure will vary with location and the shape of the coast.

What affects the rate of coastal erosion?

Some coastlines are more resistant than others to erosion by the sea. The rock type, the geological structure and the shape of the coastline are important factors in the rate of coastal erosion.

Rocks such as limestone, chalk and granite are resistant to erosion. These rocks often form cliffs and headlands. Less resistant rocks, such as clay or shale, are more easily eroded by the sea. These often form wide beach areas and low cliffs.

The geological structure can help shape the coastline. Figure 1.58 shows part of Dorset known as the Isle of Purbeck. The geology shows the rocks are aligned west to east and that the rocks have a different resistance to erosion. In the east, where the rocks outcrop at right angles to the sea there is a series of headlands and bays. This is a **discordant** coastline. In the south where the rocks are parallel to the sea, there is a **concordant** coastline, with cliffs formed by the resistant band of limestone.

FIGURE 1.59

Cliffs in Pembrokeshire

The shape of the coastline can expose or offer protection to different parts. A headland may be open to the full force of the sea, but this headland may also protect a bay. Building a harbour or other form of sea defences in the sheltered bay may disrupt natural processes. A good location has a short fetch and is sheltered from the **prevailing wind** so it does not have powerful waves.

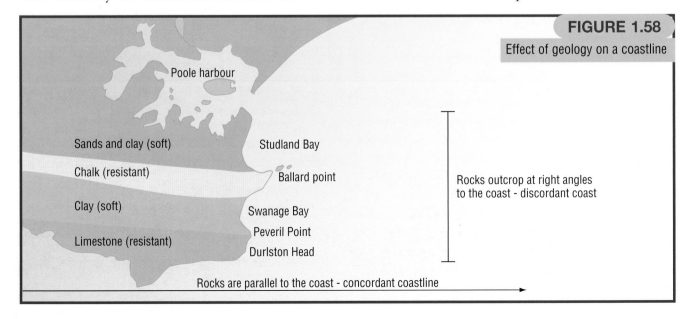

FIGURE 1.58

Effect of geology on a coastline

Poole harbour

Sands and clay (soft)

Chalk (resistant)

Clay (soft)

Limestone (resistant)

Studland Bay

Ballard point

Swanage Bay

Peveril Point

Durlston Head

Rocks outcrop at right angles to the coast - discordant coast

Rocks are parallel to the coast - concordant coastline

How the sea erodes

The main cause of **erosion** is the impact of the waves on the rock. The waves can erode in five principal ways. Often more than one process is at work (Figure 1.61).

▶ **Corrasion / Abrasion** Here the breaking waves scoop up stones and rock fragments and hurl these at the cliffs. This has the effect of chipping away at the rock, eventually breaking pieces off.

▶ **Scouring** Waves which break at the base of a cliff swirl and remove loose rock with the strong current.

▶ **Hydraulic action** The pressure exerted by breaking waves traps and compresses air in cracks. The intense pressure forces open the cracks further so weakening the rock.

▶ **Solution** Some rocks have a chemical composition which salt water can dissolve. In chalk and limestone, the calcium carbonate is dissolved, so weakening the rock.

▶ **Attrition** The waves swirl rock fragments about and as they collide they wear down. Eventually the fragments are reduced to sand or silt. Often these rounded fragments are sorted by the sea into deposits according to the size. The heaviest and largest material is left near the cliff. The sea moves, grades and deposits other material according to size away from the cliff so the finest material is deposited as sand off the coast (Figure 1.60).

As well as the sea at work, other **weathering** processes such as the wind, rain and frost also wear away the cliffs.

TASKS

1 List three factors which can affect the rate of erosion.
2 Suggest a link between the geology and how the sea has eroded the area shown in Figure 1.58.
3 Explain how each of these processes can erode the coast:
   a corrasion
   b scouring
   c hydraulic action
   d solution.
4 Describe the process which sorts beach material.

FIGURE 1.60

Grading of beach material

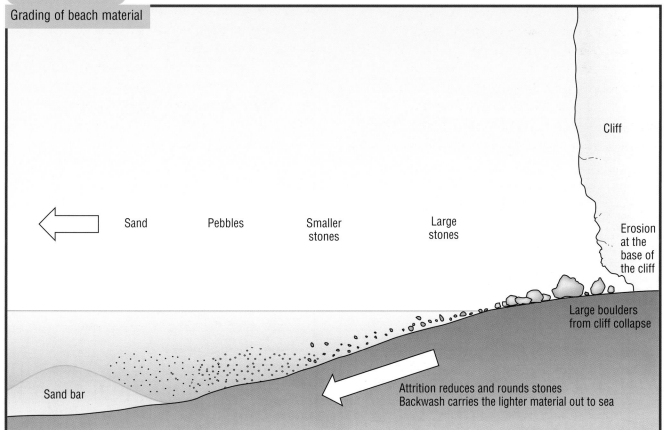

Sand   Pebbles   Smaller stones   Large stones

Cliff

Erosion at the base of the cliff

Large boulders from cliff collapse

Sand bar

Attrition reduces and rounds stones
Backwash carries the lighter material out to sea

FEATURES PRODUCED BY EROSION

The erosion of the coast produces cliffs, headlands, caves, arches, stacks and wave-cut platforms. There is a sequence to their formation, shown in Figure 1.61.

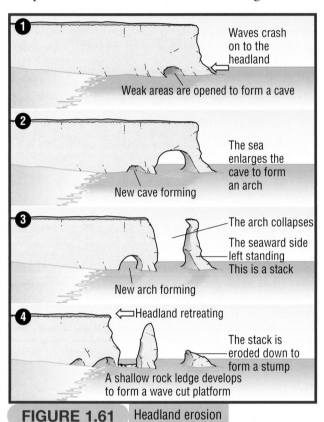

1 Waves crash on to the headland
Weak areas are opened to form a cave

2 The sea enlarges the cave to form an arch
New cave forming

3 The arch collapses
The seaward side left standing
This is a stack
New arch forming

4 Headland retreating
The stack is eroded down to form a stump
A shallow rock ledge develops to form a wave cut platform

FIGURE 1.61   Headland erosion

TASKS

1 Using Figure 1.62, describe how the different features were formed.

2 Draw a diagram to suggest what the headland would look like after Number 4.

3 Explain why slumping is only partly caused by the sea.

4 a Draw an outline of the coastline shown on the OS extract (Figure 1.63).

   b Use a key symbol and a different colour to mark these features:

   cliff   wave-cut platform

   slumping   beach

5 Using your completed map, write a commentary to describe the coastal scenery from 037900 to 100832. Start with *'At grid reference ... there is a ... . The height of the cliff here is ... m. Between grid references ... and ... there is evidence of slumping.'*

The OS extract shows a coastal area south of Scarborough where erosion is taking place. This area is composed of a resistant rock, limestone, with some thinner beds of less resistant clay. The area is covered with boulder clay and sand and gravel deposited at the end of the last Ice Age. In places the limestone has been weakened by faulting and joints. These areas of weakness are attacked by the sea. There is a wave-cut platform, Castle Rocks at 095835, a headland and cliff at Castle Cliff (052891) and bays at Cayton Bay at 071843, and North Bay and South Bay at Scarborough.

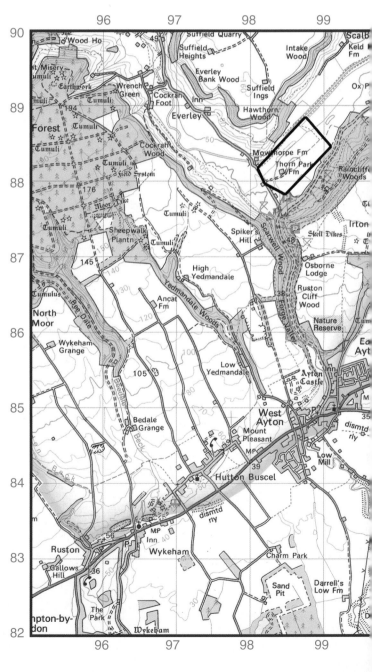

There are also several areas where the cliff has slipped, for example, Tenants Cliff (0684) and at Holbeck Hall on the South Cliffs in 1993. This is called **slumping**, caused when the sea erodes the base of the cliff leaving the rocks above unstable. The **permeable** deposits and limestone allow water to soak through. Water collects where the **impermeable** clay underlies the area. This provides a surface for the unstable rocks above to slip down (Figure 1.62).

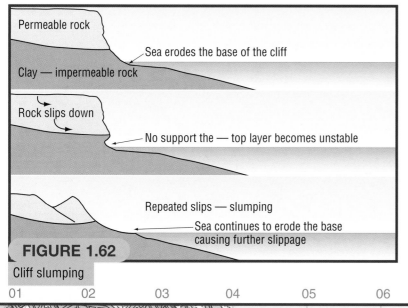

Permeable rock

Sea erodes the base of the cliff

Clay — impermeable rock

Rock slips down

No support the — top layer becomes unstable

Repeated slips — slumping

Sea continues to erode the base causing further slippage

FIGURE 1.62

Cliff slumping

FIGURE 1.63

OS Landranger Map of Scarborough

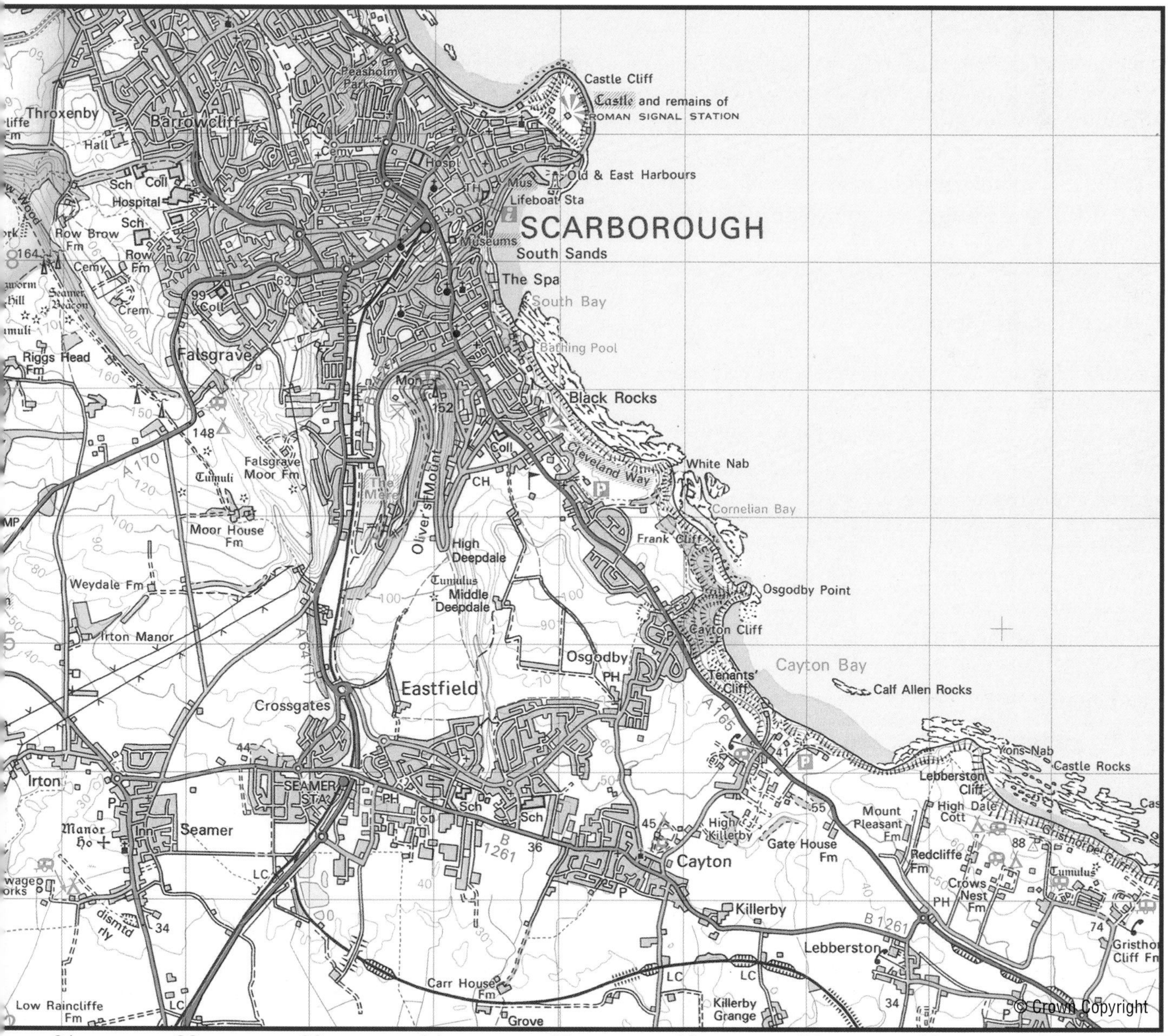

© Crown Copyright

MOVING AND DEPOSITING MATERIAL

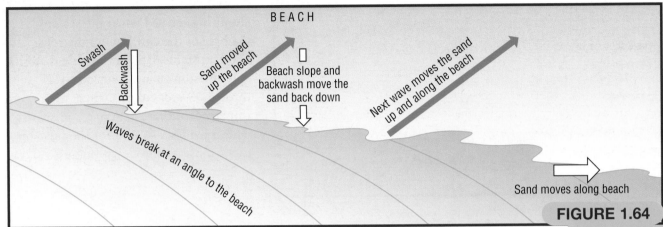

BEACH

Swash

Backwash

Waves break at an angle to the beach

Sand moved up the beach

Beach slope and backwash move the sand back down

Next wave moves the sand up and along the beach

Sand moves along beach

FIGURE 1.64

Longshore drift

The material produced by weathering and erosion is moved by waves along the coast and out to sea. This process of **longshore drift** provides the link between the wearing down and building up of the coastline. The movement of material builds up to form distinctive features including **spits**, **tombolo**, **beaches** and **bars**.

Longshore drift

Longshore drift is at work when waves break at an angle to the beach. The swash runs up the beach at an angle. However, the backwash flows straight back due to the beach's slope. A pebble being moved along has a zig-zag path (Figure 1.64).

The rate of movement depends on several factors; the strength of the waves, the size of the material being moved and the beach incline. Larger waves are able to move bigger sediment. But if there is a very gentle slope to the beach slope or the waves lack energy, little longshore drift will take place.

Moving material

Waves move material in a variety of ways. This depends on the size of the material and how much energy there is, (Figure 1.65).

▶ **Solution** Salts and minerals dissolved from the rocks.

▶ **Suspension** Finer materials can be carried in suspension by waves.

▶ **Saltation** Material which is just too heavy to be carried in suspension will be bounced along.

▶ **Traction** Heavier material is rolled along the sea bed by the strongest waves.

FIGURE 1.65

How sediment is moved by the sea

1 Fine sediment held in suspension

2 Pebbles —too heavy to be in suspension, but too light to sink are bounced along — saltation

3 Traction —heavy and large sediment rolled along the sea bed

Features produced by deposition

Where there is deposition, longshore drift provides the material for broad beaches, salt marshes, bars and spits. **Sand dunes** develop when onshore winds transport sand from the beaches to the land where it piles up as dunes.

A **spit** may form when there is a break in the coastline, for example, a river estuary. Longshore drift deposits material across at a rate faster than the river can remove it. Gradually a finger of land grows across, which vegetation may consolidate (Figure 1.68).

A **bar** is where the spit has developed across a bay or inlet creating a lagoon and salt marsh behind (Figure 1.68). Sometimes a spit may grow out from the land to connect an offshore island. When Chesil Bank joined with the Isle of Portland, it formed a **tombolo** (Figures 1.67 and 1.68).

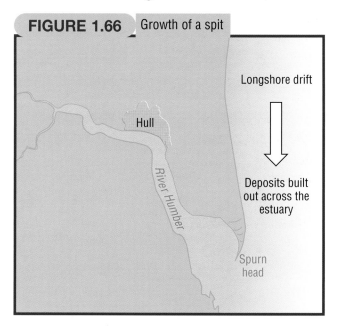

FIGURE 1.66 Growth of a spit

Hull

Longshore drift

River Humber

Deposits built out across the estuary

Spurn head

FIGURE 1.67

Chesil beach

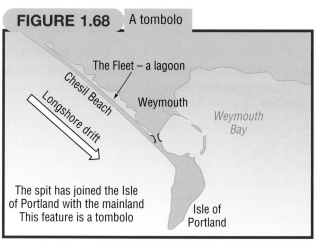

FIGURE 1.68 A tombolo

The Fleet – a lagoon

Chesil Beach

Longshore drift

Weymouth

Weymouth Bay

The spit has joined the Isle of Portland with the mainland This feature is a tombolo

Isle of Portland

TASKS

1 Describe the process of longshore drift.

2 Suggest a way you could test for longshore drift along a coast.

3 Why is there a link between the energy of the waves and how material is moved along the beach?

4 Explain how the following features were formed:

   a a spit

   b a bar

   c a tombolo.

EROSION OF THE HOLDERNESS COAST

The Holderness coast lies on the eastern side of the UK, south of Scarborough, facing the North Sea (see Figure 1.70). Most of the coastline consists of cliffs between 20 m and 30 m high, but where it meets the estuary of the River Humber an impressive spit of sand – called Spurn Head – has been formed (see Figure 1.69).

The cliffs of the Holderness coast have been eroding at a rapid rate for thousands of years. The average rate of retreat is estimated at 1.2 m per year, which is said to be the fastest in Europe.

FIGURE 1.69

Spurn Head, a spit

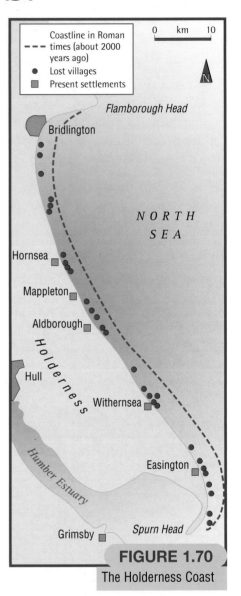

- - -	Coastline in Roman times (about 2000 years ago)
●	Lost villages
■	Present settlements

0   km   10

N

Flamborough Head

Bridlington

NORTH SEA

Hornsea

Mappleton

Aldborough

Holderness

Hull

Withernsea

Humber Estuary

Easington

Grimsby

Spurn Head

FIGURE 1.70

The Holderness Coast

Causes of rapid cliff erosion at Holderness

The area is made of a soft material called boulder clay, which is sand and clay deposited during the Ice Age. The cliffs are easily undermined by wave action and they collapse in a series of slips and slumps, especially when the material is weakened by waterlogging after heavy rain (see Figure 1.71).

The coast is frequently attacked by powerful destructive waves which are driven by strong winds blowing across the North Sea. The destructive power of the waves may be increasing because of a slow rise in sea level. The waves erode the base of the cliffs by means of hydraulic action and abrasion.

Much of the eroded material is fine clay which is carried out to sea; the amount is estimated to be 1.5 million tonnes of clay per year. The rest of the eroded material is heavier sand which settles to form a thin beach on which material is constantly moving south at a rate of about 0.5 million tonnes per year. This removal of beach material means that wide beaches cannot develop to the size where they reduce wave energy by friction; this helps to explain the powerful erosive effect of the waves on the Holderness coast.

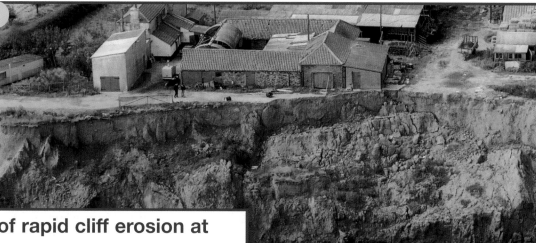

FIGURE 1.72

Sue Earle's farm, on the Holderness Coast

The effects of rapid cliff erosion at Holderness

A 4-km strip of land has been lost to the sea since Roman times, and many villages and farms have disappeared (see Figure 1.70).

The North Sea gas terminal at Easington is one of two east coast sites where gas pipelines from North Sea drilling platforms come ashore. It is a vital part of the UK's energy supply infrastructure, but it is under threat because of its cliff-top location.

FIGURE 1.73

Sea wall and groynes

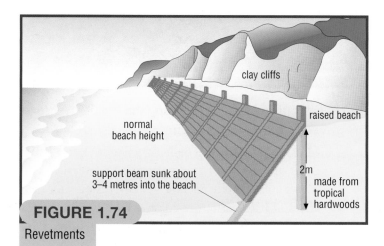

FIGURE 1.74

Revetments

clay cliffs

normal beach height

raised beach

support beam sunk about 3–4 metres into the beach

2m

made from tropical hardwoods

FIGURE 1.72

Rock barrier to protect caravan site

Coastal towns, including Withernsea and Hornsea, farms, roads and tourist facilities such as caravan sites are under threat from the advancing sea (see Figure 1.72).

Attempts to control coastal erosion at Holderness

Coastal protection measures attempt to reduce the force of waves on the cliff base by:

▶ building wave-resistant structures along the cliff base; the main types are sea walls (see left hand side of Figure 1.73) and **revetments** (sloping wooden barriers, see Figure 1.74);

▶ trapping moving beach material by means of a series of barriers – called **groynes** – which run at right angles to the cliff base (see right-hand side of Figure 1.76 and also Figure 1.77).

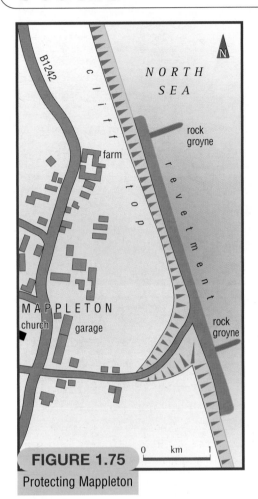

FIGURE 1.75

Protecting Mappleton

Coastal protection is expensive and arguments persist as to who should pay: local authorities with a coastline argue that coastal protection is a national issue which should be centrally funded.

Cost/benefit analyses mean that only the most valuable, populated sections of the coast receive full protection, while remote, unpopulated farmland may be left to the 'do nothing' option. Withernsea and Hornsea have expensive protection measures and the cost of protecting the Easington gas terminal with a 1 km sea wall was put at £4.5 million. In some places concrete blocks and rubble have been dumped in front of the cliff base as a temporary barrier to absorb wave energy (see Figure 1.72).

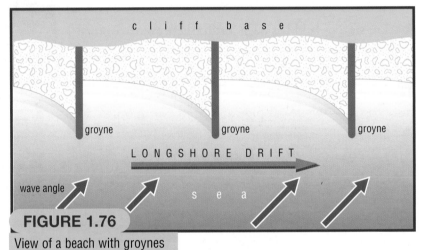

FIGURE 1.76

View of a beach with groynes

FIGURE 1.77

How the coast has eroded

By 1988 the cliff top had retreated to within 500 m of the main coast road (B1242) through the village of Mappleton. The high cost of re-aligning the road and rehousing villagers meant that a coastal protection plan was drawn up (see Figure 1.75) and a rock barrier was constructed in 1991.

Other proposals for protecting the Holderness coast involve constructing submerged banks out at sea:

▸ One idea is to dump colliery waste 50 km off the coast, in a series of banks designed to refract waves and thereby encourage the erosion of a series of bays along the coast, which would then hold back protective sand against the tendency for longshore drift. This would help to solve the

FIGURE 1.78
Wave damage to the road along Spurn Head

problem of disposing of the millions of tonnes of waste produced by coal mining each year. Experience of dumping colliery waste in the sea off Northumberland has shown that it forms hard, stable banks but it makes the sea dirty and turns beaches black.

▶ Another idea is to submerge a bank of 1.5 million compressed tyres, bound together with a mesh of nylon ropes and concrete. The resulting bank would be located 1 km offshore and would be 6 m high, 110 m long and 60 m wide. The bank was proposed as a trial which – if successful – would be followed by a full scheme involving more than a billion tyres arranged in seven, 2-km strips along the coast.

The possible 'knock-on' effects of coastal protection at Holderness

Inhabitants of the coastline south of Mappleton are convinced that the construction of the rock groyne (see Figure 1.75) in 1991 has accelerated erosion of their cliffs to a rate of 10 m per year by blocking the southward movement of protective sand. Farmers Sue Earl and her uncle were forced to abandon fields, outbuildings and – eventually, in 1996 – their farmhouse as the cliffs retreated at an alarming rate (see Figure 1.71).

It is possible that the balance of Spurn Head as a natural system has been disrupted by measures designed to protect it from erosion. Victorian groynes have encouraged the formation of a dune belt near the spit's tip and concrete sea defences at its northern end have exaggerated its exposure as the cliffs to the north have retreated westwards. The spit – and the road which runs its length to the isolated homes of the Humber lifeboat crew – may be exposed to more severe erosion (see Figure 1.78).

A study at Hull University recommended no further intervention and allowing nature to run its course. Taking this attractive 'low cost' option could mean that waves would erode Spurn Head on its eastward side and deposit on its westward side, enabling the spit to move bodily westwards at about 2 m per year, keeping pace with the rest of the Holderness coast.

These two examples of 'knock-on' effects show why it is important to try to understand the natural systems we may wish to intervene in; the complexities of the Holderness coast system as a whole may be part of an even bigger, more complex system which is moving material around the whole of the North Sea basin. Local intervention may have international effects.

COURSEWORK FOR RIVERS

Coursework based upon fieldwork along rivers is the most common type of physical geography enquiry, mainly because most people live within easy reach of a river of suitable size (Figure 1.79). River based studies can be entirely physical, but they don't have to be as there are plentiful opportunities to develop human themes as well.

FIGURE 1.79

Measuring channel width. This is probably the largest sized river in which it is safe to take measurements.

River studies are good for:

▶ collection of data in a group;

▶ people who prefer data collection by measurement and observation;

▶ collecting large amounts of data for processing and presentation.

Suggested titles

The title can be quite general.

▶ A study of the characteristics of stream A.

▶ What are the main features of stream B?

General titles such as these give a lot of freedom, and references to both physical and human geography can be included in varying proportions. Also, if the data is going to be collected by a group, each member of the group can concentrate upon a different aspect when they write up the work. If the stream is a short one (say under 10 km long), it may be possible to study its full length by selecting a number of study sites along it.

▶ A study of river C and its valley from source to mouth.

▶ How does river D and its surroundings change from source to mouth?

▶ In what ways and why do channel characteristics of river E change as you travel from source to mouth?

Many rivers and some streams are too long for a complete study; a small section needs to be carefully chosen. A section along which there are marked differences in river features is best because this will generate more comment. Differences are stressed in some titles.

▶ What are the differences in channel features above and below point F?

▶ Are the channel features of river G different between winter and summer?

▶ Are there any differences in river pollution above and below settlement H?

Some people prefer to make a special study of just one aspect of the geography of a river.

▶ Where and why has river I deposited its load?

▶ What evidence is there that river J is polluted?

▶ A study of flood prevention measures along river K.

▶ What are the causes and consequences of flooding by river L?

The coursework process for a river study

There are six stages in the process.

1 Find a title

Think of a working title. You can always change it later if it proves to be unsuitable.

'Is river R a typical river?' is one example which could be used as a starting point.

> **Advice**
>
> The title must clearly indicate the main aim of your study.

2 Think about the practical possibilities

Which river? Is there a river of suitable size not too far away?

Where can the data be collected along this river? Is there good public access to the river?

> **Advice**
>
> Ask yourself: 'Is it going to be possible to collect sufficient data for a successful enquiry?'.

3 Plan the work

▶ Look first at the course of the river on a large-scale map, such as an OS map, and select some possible sites for data collection. Five or six would be a typical number of sites.

▶ Devise three or four smaller hypotheses to support the main aim.

▶ Decide methods of data collection to be used at the chosen sites (see pages 54–55) .

▶ Get ready for the data collection by assembling the equipment needed and making your data recording sheets.

> **Advice**
>
> Title: 'Is river R a typical river?'
> Examples of sub-hypotheses:
> As you move downstream, the river will have
>
> ▶ a deeper and wider channel
>
> ▶ a greater discharge
>
> ▶ an increase in velocity
>
> ▶ a less steep gradient
>
> ▶ boulders smaller in size.

4 Undertake the data collection

Measure and observe as carefully and accurately as possible. Carry out the work in the same way at each site so that results can be reliably compared. Often you have only one opportunity to collect data so gather as much as you can in the time available.

> **Advice**
>
> To help you memorise what it was like when you come to write it up later, draw labelled field sketches or take photographs, or do both.

5 Process and present the data collected

After taking river measurements, some calculations can be done, such as cross-sectional area and discharge for each site, or average values for surface speed of flow and size of bedload at individual sites. Techniques of presentation most commonly used in river studies are maps, data tables, channel cross-sections, line and bar graphs, labelled field sketches and photographs.

> **Advice**
>
> Look for a range of relevant map and graphic techniques; about six different and appropriate types are needed.

6 Write up the work

You need at least four sections:

▶ Introduction

▶ Data Collection

▶ Analysis and Interpretation of the Results

▶ Conclusion.

Methods of measurement

Figure 1.80 illustrates some of the measurements that can be made at each site along the stream.

FIGURE 1.80

Site measurements along a river

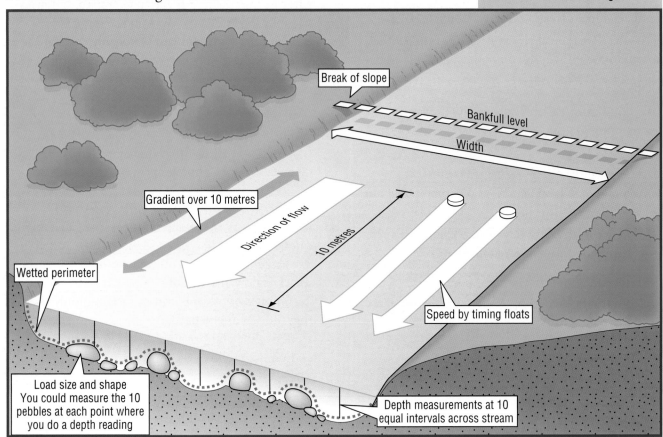

Break of slope

Bankfull level

Width

Gradient over 10 metres

Direction of flow

10 metres

Wetted perimeter

Speed by timing floats

Load size and shape
You could measure the 10 pebbles at each point where you do a depth reading

Depth measurements at 10 equal intervals across stream

Width Use a tape or rope to measure the width at the water surface from bank to bank.

Depth At regular intervals across the channel measure the depth of water from surface to stream bed with a metre rule.

Wetted perimeter Lay a rope or long chain on the stream bed. Make sure that it touches the stream bed. It can be quite a wet job getting it to lie on the bed up and over rocks and stones.

Velocity (speed of flow) One of the easiest ways to measure surface speed is to time how long it takes a float (such as piece of wood, orange or onion) to cover a 10-m stretch of river. For increased accuracy, do this several times at three or more places across the stream and take the average of the results. Surface speed at any point is distance divided by time. The surface flow is slightly slowed due to friction with the air above so you can multiply the

result by 0.85 which gives a more accurate idea of the stream's overall velocity. If a flow meter is available, you can make a more detailed study of velocity and take readings at different depths and in different positions across the river.

Gradient At least two people are needed to hold the poles or metre rules and take a reading from the clinometer; it is best to have a third to do the recording (Figures 1.81 and 1.82).

Load size Try to collect the pebbles to be measured in as random (unbiased) a way as possible. You need to think carefully about how this can be done. For example, at the same location you take the depth reading, you could use the pebble that the metre rule touches and another nine around it. You can measure each pebble's long axis (a) or width (b) and take the average for the location (Figure 1.83).

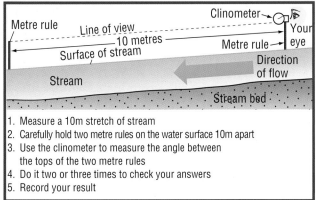

FIGURE 1.81

Measuring a stream's gradient

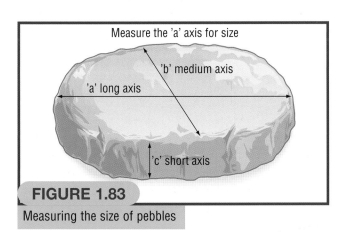

FIGURE 1.82 How to measure gradient

1. Measure a 10m stretch of stream
2. Carefully hold two metre rules on the water surface 10m apart
3. Use the clinometer to measure the angle between the tops of the two metre rules
4. Do it two or three times to check your answers
5. Record your result

FIGURE 1.83

Measuring the size of pebbles

Measure the 'a' axis for size

'b' medium axis
'a' long axis
'c' short axis

Collecting data

Advice

▶ Take safety seriously
▶ Wear Wellington boots
▶ Do not make channel measurements when the river is in flood
▶ Always have someone there to help you with the measurements
▶ Wear warm clothing
▶ Pick the right time for doing the fieldwork
▶ A fine day without too much wind makes measuring and recording easier
▶ Try to do the fieldwork after rainfall when there is a sufficient flow of water
▶ Do all the work in one or two days so that readings from different sites can be compared
▶ Don't go near a stream with bankfull or flood conditions

FIGURE 1.84

This is only a small stream, but high water levels make it too dangerous for taking measurements.

Make an equipment check list similar to the one below:

Equipment checklist

- OS or large-scale map of area of study
- Recording sheets on a clipboard
- Notepad, pens and pencils
- Tape measure and long rope or chain
- Metre rules
- Poles and clinometer (for gradient)
- Floats and stop watch and/or flow meter
- Wellingtons, sweater, waterproofs, drinks and lunch
- Camera and film

Make a separate recording sheet for each site. Include spaces only for the data that you are intending to collect. Figure 1.85 is an example of a recording sheet which should be adapted for the nature of the data being collected. Remember to leave spaces for recording essential details about each site. You should also make brief notes on the location and its main features in your notebook before beginning to take measurements at each site.

Site number						Grid reference		
Date		Weather						
Notes about the site								
Width		Gradient						
Depth (cm) (m from bank)	0.5	1.0	1.5	2.0	2.5	3.0	3.5	4.0
Float times (seconds) (m from bank)		1.0		2.0		3.0		
Stone long axis (cm) Stones	1		2		3		4	5
at 1.0m from bank								
at 2.0m from bank								
at 3.0m from bank								

FIGURE 1.85

Example of a recording sheet for rivers

Processing and presenting the data collected

The first task is to take the values from your separate recording sheets and put them into tables of data such as the one shown in Figure 1.86. Once clearly tabulated, it is easier to process and present the results.

The values from Figure 1.86 can then be used to draw channel cross-sections for each site. Two of these for Sites 5 and 6 are shown in Figure 1.87. It is important to keep the same scale at each site so that they can be compared.

Site	1	2	3	4	5	6	7	8
River width (m)	1.7	2.7	5.2	4.3	5.2	7.1	8.2	7.0
Depth (cm) at:								
0.5m	20	10	10	10	10	6	5	12
1m	25	15	10	10	12	7	6	10
1.5m	15	20	11	11	15	7	6	10
2m		10	12	12	17	8	5	10
2.5m		10	14	12	22	8	4	9
3m			15	10	27	8	4	10
3.5m			15	11	28	10	4	10
4m			13	10	25	15	7	11
4.5m			12		20	17	9	14
5m			10		7	20	12	16
5.5m						19	13	20
6m						19	16	23
6.5m						18	18	17
7m						14	20	0
7.5m							24	
8m							20	

FIGURE 1.86

River measurements at eight sites

By tracing the outlines of these cross-sections onto graph paper, it is possible to work out the area of the channel at each site. If river velocity at each site has been measured, the discharge at each site can be calculated by using the formula:

$$Q = A \times V$$

where Q is the discharge (m³/sec or cumecs)

A is the cross-sectional area (m²)

V is the velocity (m/sec)

A bar graph can be used to show the size of the discharge at each site with the expectation that the bars will increase in length further downstream.

FIGURE 1.87

Channel cross-sections

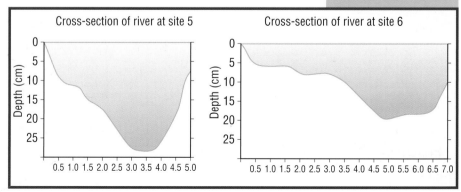

The next task is to assemble and organise the visual materials – sketches and photographs. Decide which of them are useful and think about how they may be used to support the aim of the work. Figure 1.89 shows how good annotation increases a photograph's impact.

Advice

- Be selective – it isn't necessary to use every photograph.
- Place them next to the information about the site at which they were taken.
- Don't put them in a block either in the main work or an appendix.
- Add labels to highlight what they show that is significant to your study.

FIGURE 1.88

Drainage basin of the river used in the study

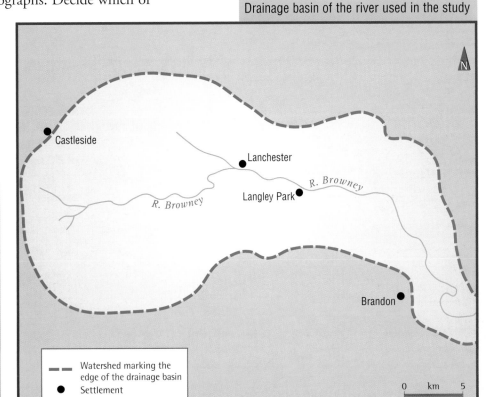

Castleside

Lanchester

R. Browney

R. Browney

Langley Park

Brandon

- – – Watershed marking the edge of the drainage basin
- ● Settlement

0   km   5

Small meander

Bank eroded

Steep bank

Gentle valley sides

Deposition on the edge of the channel

Slip off slope forming

Riffles

FIGURE 1.89

A stream in its middle course.

FIGURE 1.90

Locations of the study sites

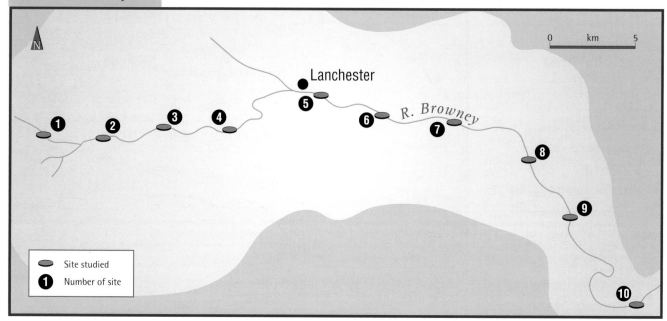

Writing it up

There are four main parts.

1 *Introduction*

> **CHECKLIST for what should be included in the Introduction**
> ▸ A clearly stated title
> ▸ Main aims of the study explained
> ▸ Information given about the study area
> ▸ Map showing its location and sites studied
> ▸ Geographical background to the topic.

There is a temptation with a river study just to show the location of that stretch of river with the study sites. It is best to show all of the river's drainage basin, as in Figure 1.88, and then use a second map on which to show site locations (Figure 1.90). Although more credit will be given if you draw your own sketch map, it is acceptable to use a printed map, provided that it is labelled and the features significant to the study are highlighted. Just enclosing a photocopy of a map is worthless because most of what is shown on it has nothing to do with the work. If you are going to answer the question 'Is river R a typical river?', it is clearly essential that you state and comment on the features of a typical river. You should refer to the expected changes in a river between source and mouth and it is wise to mention, at least briefly, some of the processes responsible. The introduction is the place to do this, but keep it short. You can cover all that is needed for the geographical background on one side of A4 paper at the most.

2 *Data Collection*

This section is likely to be longer than the Introduction. You must remember that this and the Introduction are together worth 40% of the marks. You cannot afford to miss out.

> **CHECKLIST for what should be included in Data Collection**
> ▸ Methods of data collection described briefly
> ▸ Methods of data collection explained
> ▸ Places where the data was collected stated
> ▸ Explanation given why these places were suitable
> ▸ Dates and times of data collection given
> ▸ Dates and times of data collection explained.

Students are usually more willing to describe methods of data collection, often with supporting sketches and illustrations, than they are to explain *why* these methods were used. They state where the data was collected, but often fail to explain *why* these places were chosen and were suitable. Frequently dates and times are not given, but these can have an important effect upon the reliability of the data collected.

> **Advice**
> For data collection, always explain
> ▸ Which methods?   Why?
> ▸ Where used?   Why?
> ▸ When used?   Why?

3 Analysis and Interpretation of the Results

Have your tables of data, maps, photographs and graphs ready before you begin to write about what the collected data shows. Remember that it is best to insert each one as near as possible to the point where you are writing about them. Write down your title and aims on a separate sheet of paper and keep it in front of you so that you can't forget what you are supposed to be writing about.

> **CHECKLIST for what should be included in Analysis of the Results**
> ▸ Is there a description of what the data and figures (maps, photographs and graphs) show?
> ▸ Are there attempts to give reasons for what is shown?
> ▸ Are there comments which relate back to the main aim of the study?
> ▸ Has the analysis been extended to include similarities or differences or relationships between data and sites?

Is the analysis well organised, with figures placed next to the written part to which they refer? The most successful analyses are those in which students show that they are thinking about the main aim of their study all the time and are always trying to give reasons for everything.

4 Conclusion

This is the chance for you to look back at all the work done. You need an overall conclusion which fits the title on the front cover. Don't worry if some of your data does not support the aims – that is only to be expected. A stream is likely to have many of the features of a typical river, but not all of them. A photograph can be a good way to show this (Figure 1.91). The conclusion gives you another chance to refer to the broader geography of rivers, as well as an opportunity to evaluate what worked well and what didn't.

Not typical of an upper course:

gentle slopes and flat land on valley floor

Typical of an upper course;

shallow water

boulders in the bed

FIGURE 1.91

River in its upper course

COURSEWORK FOR COASTS

The coastal scenery of the British Isles is noted for the variety of its landforms (Figure 1.92). The coastline is also an attractive environment for people, both residents and visitors, which leads to the need for coastal protection (Figure 1.93). This means that all around the coastline there are plentiful opportunities for undertaking coursework which is entirely physical, or mainly human, or a mixture of the two.

FIGURE 1.93

Coastal protection at Eastbourne

FIGURE 1.92

Off Flamborough Head. Which landforms can you recognise?

Suggested titles

A. Physical

▶ An investigation of the landscape features of a stretch of coastline and the reasons for them.

▶ What are the similarities and differences between two stretches of coastline?

▶ What are the features of the beach and cliffs along a stretch of coastline and how were they formed?

▶ How and why do beach profiles change either along one stretch of coast or between summer and winter?

B. Mainly human

▶ An investigation into the methods of coastal protection along a stretch of coast and the reasons for them.

▶ What are the similarities and differences between the ways in which people use two stretches of coastline?

▶ Why is one part of the coastline more popular with visitors than another?

▶ What problems are caused by people living along and visiting a stretch of coastline?

C. Mixed physical and human

▶ An investigation into the physical features and human uses of the beach at X.

▶ What are the natural and human features of a stretch of coastline?

▶ What is the evidence of coastal erosion and what methods are used to reduce it?

▶ What are the impacts of people upon the beach at Y?

Advice

Avoid titles such as: 'How successful have attempts been to control coastal erosion? It is difficult to gauge how successful anything has been. The data collected is unlikely to lead to an answer.

Methods of data collection

Whatever your preferred method(s) of data collection – measurement or observation or questionnaire – opportunities exist in coastal fieldwork. There are also several sources of secondary data which provide supporting information.

Primary data collection

▶ Measuring cliff profiles
▶ Measuring beach profiles
▶ Measuring pebble sizes
▶ Observing features of coastal erosion
▶ Observing features of coastal deposition
▶ Observing methods of coastal protection
▶ Questioning local residents and visitors

FIGURE 1.95

Shingle beach showing changes in slope downshore

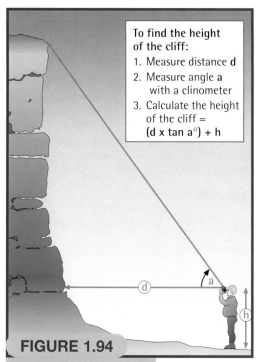

To find the height of the cliff:
1. Measure distance **d**
2. Measure angle **a** with a clinometer
3. Calculate the height of the cliff = **(d x tan a°) + h**

FIGURE 1.94

Finding the height of a cliff

Secondary data collection

▶ OS and geology maps
▶ Visitor guides
▶ Council visitor surveys
▶ Information on coastal defences
▶ Newspaper reports

Cliff profiles

1 Start by drawing a sketch of the cliff. You could back this up with a photograph.

2 Estimate the height of the cliff using the method shown in Figure 1.94. Check your result against the contours on an OS map.

3 Indicate on the sketch where the slope angle changes and any smaller scale features such as joints, bedding planes and caves, as well as vegetation and signs of landslips. Check with the geology map.

4 Use the secondary sources to help you explain the shape and other features of the cliff profile.

Beach surveys

Measurements are normally taken along a line, or transect, down the beach. The downshore beach profile will show changes in slope. There may also be changes in pebble size as well.

1 Take a general clinometer reading from the edge of the cliff to the low water mark.

2 Record the angle of slope at intervals down the beach. Often it is best to select the breaks of slope, which can be clearly seen on the shingle beach in Figure 1.95.

3 At regular intervals down the beach, randomly select 5 to 10 pebbles to measure the long axis and shape. This can be done by placing quadrats on the beach. It may be desirable to do more transects further along the beach. Any changes in profile and pebble size are likely to result from the work of longshore drift. If you are going to question users of the beach, two questionnaires may be better than one – one for locals and one for visitors.

EXAMINATION TECHNIQUE – UNIT 1

Now that you have learned about 'People and the Physical World' it is time to investigate how you can use your knowledge and understanding to answer GCSE questions.

All GCSE questions use geographical resources. There is aways at least one question which focuses on a photograph. There is only a small chance that you will have seen the photograph before so the questions begin with shorter sections to get you used to the photograph and question topic.

The following question is from the Foundation Tier (Paper 1) and is based on a photograph.

> **1 (a)** Study Photograph B which shows a river near its source.

Photograph B

The **question theme** is obviously rivers, but on careful reading the theme can be sub-divided into

1 river features,

2 river flooding.

> **(i)** What is the source of a river? [1]

The starter question tests your knowledge of geographical **terminology**. It is a good idea to compile a **glossary** of geographical words which will include 'source'. There is only 1 mark allocated to this question so a definition or concise description is needed.

> **(ii)** Describe **three** natural features of the river and its valley. [3]

You must study the photograph in order to answer this question. When you have identified three different features from the photograph they must be described. Write in full sentences, do not use one word answers when you are asked to describe.

> **(b)** Study Fig. 2 below.

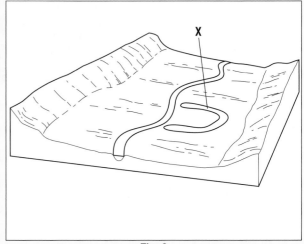

Fig. 2

> **(i)** Name the feature labelled X on Fig. 2. [1]

The **command word** 'name' instructs you to answer briefly.

> **(ii)** On Fig. 2 label the flood plain. [1]

Questions b(i) and b(ii) are testing your knowledge of river features.

> **(c)** For a named river which has flooded:
> – explain why this river flooded,
> – describe the effects of the flood. [6]

The final question has half the total marks and tests your knowledge of a **case study**. You may choose any river which you have studied that has flooded. This might be a river in your local area or a major river of the world such as the Mississippi. There will be one mark reserved for the name of the river, so if you cannot name the river you may still score some marks but will not get full marks.

The question tests your knowledge of why the river flooded and your understanding of how the flooding affected people and the area around the river. There are 5 marks for this, divided between the two parts of the answer so there will be a maximum allocation of 3 marks each for either the causes or effects. The other 2 marks must be scored on the other topic.

The following question is taken from the Higher Tier (Paper 2). Notice how both questions use the same photograph resource.

> **2 (a)** Study Photograph B. Give two pieces of evidence which show that this is an upland river valley. [2]

This question must be answered by using evidence from the photograph.

> **(b)** Describe the processes by which a river erodes the valley in an upland area. [4]
>
> **(c)** Explain how river processes may lead to the formation of an ox bow lake. You may draw a labelled diagram as part of your answer. [5]

These questions require detailed knowledge and understanding of river processes. Words such as **corrasion** and **hydraulic action** can be used to show your knowledge of geographical terms. The use of such words from your glossary will enable you to give a sufficiently detailed answer. Notice how question (c) gives you the opportunity to draw and label a diagram in your answer as an alternative to writing prose. However, as the question says 'you may draw a diagram', you will not lose marks if you do not include one.

> **(d)** For a named river which has flooded:
>
> **(i)** explain why this river flooded [4]
>
> **(ii)** describe the effects of the flood on people and the environment. [5]

As in Paper 1, the case study question is allocated most marks, almost half the total for the question. A detailed answer is required that obeys the different command words. Your answer should include specific facts about the causes and effects of a flood. One mark is again reserved for correctly naming a river which must relate to both parts of the question.

The main difference between these questions from the Foundation and Higher tier papers is the amount of detail which is required in the answer. Paper 2 needs more extended writing, whereas answers to Paper 1 are likely to be shorter and more structured under separate topic headings.

The following question is taken from the Higher Tier paper.

> **3 (a)** Describe methods that can be used to protect the coastline from erosion. You should refer to one or more areas you have studied. [7]

The question provides an opportunity to use case study knowledge. It focuses on situations where coastlines are being protected from the erosive power of the sea. The answers below show how three candidates have had different amounts of success in tackling this question.

Answer from candidate 1

At Barton On Sea the cliff face was being rapidly eroded. This had to be stopped as there were three landmarks on top of the cliff: a golf course, Highcliff Hotel, and a small village. The local authority decided to build groynes to keep beach material in place, and to stop the erosion of the cliff they put concrete posts at the foot of the cliff with a trench dug behind them which was filled with boulders. This was done to prevent the beach material being hurled against the cliff face eroding the base. A thick metal sheet was inserted half-way up the cliff to try to hold it together and stop it disintegrating. Also drains were put into the cliff to stop water building up in it. This made the cliff more stable.

Examiner's comment

The candidate begins in the best way possible by naming the area which has been studied. The rest of the answer then clearly refers to this chosen area. The answer is put into context by some background information being provided about why the coastline needed protecting. Useful though this information is, no marks are scored because it does not answer the question. The answer then goes on to describe five different methods of protection which have been used on this coastline. Not only are these methods described but it is also explained how they work successfully. This explanation is credited as the 'development' of ideas and scores extra marks.

Answer from candidate 2

In 1987 the beach at Seaford on the south coast of England was properly defended against erosion and to stop flooding. First beach material was dredged from the port at Newhaven nearby, then it was deposited on Seaford beach. Groynes were built perpendicular to the coast to try to stop longshore drift or slow it down. To save all the dredging concrete sea defences can be made.

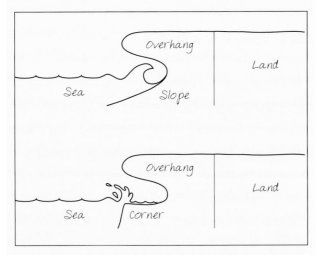

On diagram A there is a slope to dissipate wave energy and in diagram B there is a corner to take the force of the waves. Both have overhangs to hold down the spray of the waves. Concrete boulders were used on Seaford beach further east to protect the chalk cliffs.

Examiner's comment

The candidate begins by naming the case study example and explains the need for protection. The first method referred to is beach feeding but the idea could have been developed by explaining how this would need to be done continuously. The candidate then mentions groynes but does not describe what they are or how they could be used with beach feeding to prevent loss of material. Longshore drift is also referred to in the answer but the important link is not made between building up the beach and preventing this new material from being removed.

The candidate then describes an alternative method of coastal protection but it is unclear whether the concrete sea defences are used to protect the beach at Seaford. The description of how the sea walls prevent erosion is well supported by the diagrams. Finally the candidate mentions another way of protecting the coastline but now relates this to cliff protection and does not explain how the concrete boulders are used.

Overall this answer shows some evidence of the candidate having learned a case study but the answer contains ideas that should have been further developed. It is also unclear if the different methods of protection are all used in the same area.

Answer from candidate 3

Placing concrete blocks at the foot of cliffs reduces erosion. The sea then breaks on the concrete blocks reducing erosion. Sea walls force the sea back onto itself and the ramp slows down the sea before it reaches the wall. Harbour walls protect bays and beaches from the force of the waves. Groynes on the beaches help to prevent longshore drift. The beach material is then kept in each section.

Examiner's comment

This is a very general answer which does not refer to any particular location. Three ideas are referred to and there is some explanation of how they can prevent erosion, but these ideas are not developed. Overall the answer is vague and includes some repetition about sea walls.

ACTIVITIES

The following examination questions are all linked to the theme of plate tectonics.

(a) Define the term tectonic plate. [1]

(b) Name any TWO adjoining plates and describe the movement along the plate margin where they meet. [1]

(c) Use the following information to plot on Fig. 1a an earthquake which occurred in Mexico in 1985.

Intensity measured on Richter scale   8.1
Number of deaths   9500
[1]

Fig. 16 (make a copy of the scatter graph)

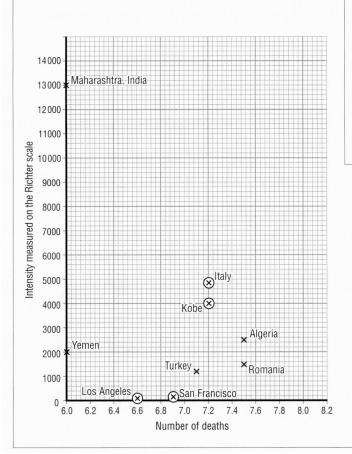

(d) Identify the earthquake which measured the same intensity on the Richter scale as the one in India but caused fewer deaths. [1]

(e) Give three reasons why many people continue to live in cities which are at risk from earthquakes. [3]

(f) Explain why many people live in areas affected by volcanic activity. [4]

(g) With reference to a named volcano you have studied describe the hazards to people and the environment caused by the volcanic eruption. [6]

(h) Describe the location of the world's main earthquake zones. Using examples which you have studied give reasons why earthquakes occur in these zones. You should develop your answer fully. [7]

Now attempt the following **tasks**.

1 Make a list of the different **command** words (page 232–33 will help you to do this).

2 Make a list of the **geographical** words. Find the meanings of any words you do not know.

3 Look at the number of **marks** allocated to each question. What do you notice about how the marks vary? Why does this happen?

4 Discuss with a partner whether each question is testing **knowledge**, **understanding**, the application of understanding or **skills** (a question may test more than one of these).

5 Two questions require you to use an **example** or **case study**. With your partner decide which examples you would use and discuss the main ideas you would include in your answer.

THE ENTRY LEVEL CERTIFICATE

If you choose to enter for the Entry Level Certificate, rather than GCSE, you will study the same Units, but not in the same depth. On this page, you can see the sort of activities for Entry Level.

People and the physical environment – rivers

Rivers flow from the mountains to the sea. As they do, they pass farms, factories and people living in towns and cities. A river changes the land and is used by people in different ways along its journey.

In the mountains, when the river starts – the **source** – it is fast and powerful. It wears away the land. This is called **erosion**. Over long periods of time, this makes channels or valleys through the land which help to move the water more easily. As the river gets nearer to the sea, these valleys become deeper and wider.

The rock that is worn away does not stay in one place. It is moved by the river. This is called **transportation**. Sometimes strong rivers move large rocks and boulders. Small rivers, which are less powerful will only be able to move small particles of soil and mud. Over many years, much of this material – the **load** – is moved downstream.

As rivers get nearer the sea and lose some of their power, they cannot carry as much of the load as they used to. The heaviest rocks are put down first. This is called **deposition**.

TASKS

1 a Copy the map opposite.
  b In the correct places, write *source* and *mouth* on your map.
2 The work of a river is *eroding*, *transporting* and *depositing*.
   Write a sentence for each to explain what it means.
3 On your map, write *eroding*, *transporting* and *depositing* where the river is doing it.

The smaller particles of sand and mud will be deposited as the river gets near the end of its journey – the **mouth** or **estuary** – and finally flows into the sea.

Rivers are very important to people. We have used them for thousands of years to carry goods by boat. Many settlements are next to rivers so that people have a supply of water every day. Over many years, our use of water and of rivers has changed. Today, we use lots of water in factories and use rivers for enjoyment.

TASKS

4 Why were many settlements built near to rivers? Write down two reasons.
5 Draw a picture to show the ways that people use rivers. Write labels on your picture.

As rivers flow across lowlands, getting nearer to the sea, they lose a lot of their power. They develop large bends called **meanders**. These meanders move slowly down the valley and across it . This makes the valley a lot wider and flatter than before. This part of the river valley is called the **flood plain**.

The land is flat, so it is easy to build on but, when the river level rises, this area is easily flooded. Homes, shops and factories would be under water. In some parts of the world, when a river floods it can kill thousands of people and cause millions of pounds worth of damage.

TASKS

6 What problems can a river cause when it floods?
7 What can people do to stop a river from flooding?
8 Look at the information on pages 32–39 about the flooding in the Rhine valley or Bangladesh.
   Choose one of them. Do a newspaper front page about it. You could produce it on computer.

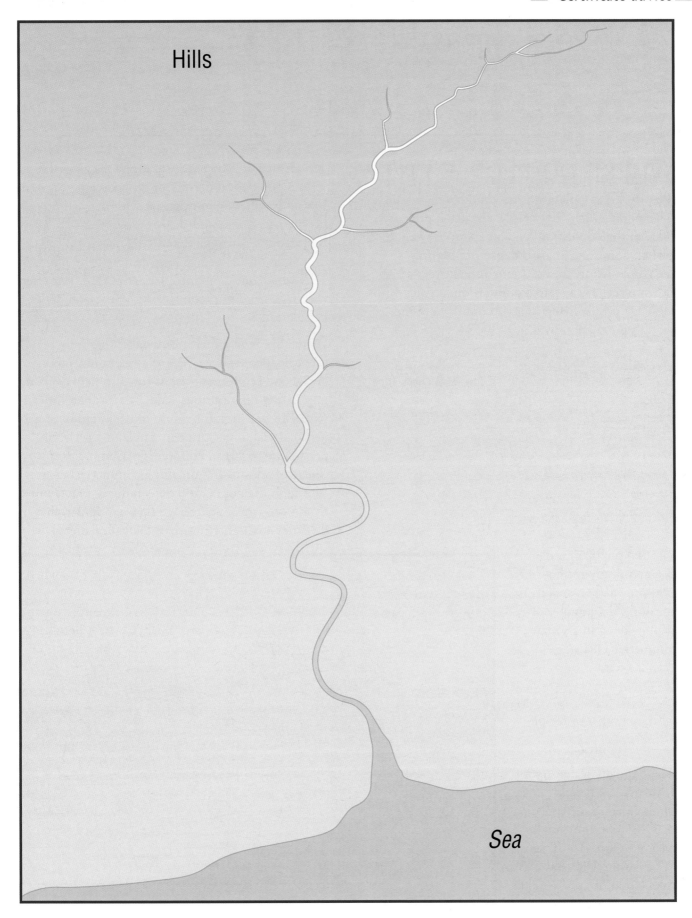

Hills

Sea

People and Places to Live

Population

WHERE DO PEOPLE LIVE?

The world's population is now more than 6 billion. This population is not evenly spread. At present 50% of the population lives on about 5% of the planet. Some areas have a **dense population**, a large population crowded into a small area, while other areas have a **sparse population**, an area with very few people living there. About 80% of the world's population lives in LEDCs (Figure 2.1).

Region	Population density (per sq. km)	Percentage of the world's population
Africa	14	10
North America	11	6
South America	16	8
Asia	84	59
Europe/CIS	39	16
Oceania	3	1

Figure 2.1 Where are the people?

Plotting population density is a useful way to show how crowded an area is. It measures the number of people living in a square kilometre. This is calculated as:

population density =

$$\frac{\text{the number of people}}{\text{area}}$$

The population density for China is 127 people per sq. km. China has a population of 1 185 000 000 and an area of 9 326 410 sq. km. Whereas the USA has a population of 263 563 000 and an area of 7 665 222 sq. km. This gives the USA a population density of 28 people per sq. km. Macau has the highest density with over 28 500 people per sq. km and the lowest are Canada 3, Botswana 3, Australia 2 and Greenland 0.2 (Figure 2.2). The world average is 46 people per sq. km. This works out as about 2 ha per person. However, within each country are areas of sparse and dense populations.

People live where their needs can be met; for example, people live where they can grow crops. Social and economic reasons can also help explain population distributions – desert areas like Kuwait with 148 per sq. km because of exploitation of a raw material, oil; south east England with 4 500 per sq. km. because London is a capital city, a centre of administration and a commercial centre. Although the population density for a country or region only gives an average figure, population densities are higher in urban than rural areas.

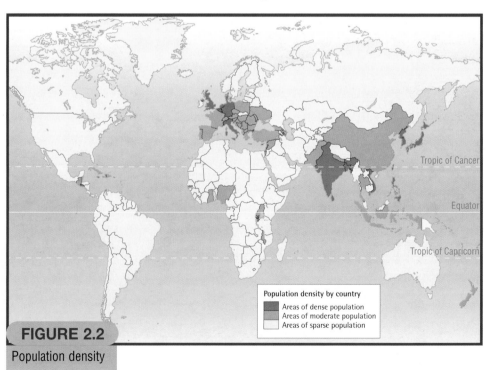

FIGURE 2.2
Population density

Population density by country
- Areas of dense population
- Areas of moderate population
- Areas of sparse population

| 70% | of the planet is ocean |
| 30% | of the planet is land |

29%	too dry
17%	too cold
11%	too mountainous
9%	infertile soil
23%	could be cultivated for food
11%	is being cultivated for food

Figure 2.3 Living on planet earth

The environment provides some reasons why people choose to live in certain areas. As Figure 2.3 suggests, only 34% of the land mass is well suited for people to live on. 45% of the economically active population is employed in agriculture. Arable land accounts for 11% of the world's land use. Fertile areas, such as river flood plains and deltas, and areas of suitable climate (not too hot or cold, with sufficient rainfall), will attract farmers and their families. Many of these farmers live in LEDCs where the percentage employed in farming is high, for example, 67% of the population of Bangladesh. These farmers are concentrated in the fertile Ganges Delta and alongside its tributaries. Figure 2.4 outlines why difficult areas like hot deserts are unattractive to farmers and have a sparse population.

Social and economic factors can help explain why urban areas attract large numbers of people. This is especially true for MEDCs where there are large numbers employed in manufacturing and service industries in urban areas. For example, the UK has an urban population of 89% and 98% of the economically active population employed in either manufacturing or services. By comparison, Bangladesh has an urban population of 17%.

Climate:
        Too hot with average temperature above 30°C
        Too dry with annual rainfall below 200 mm
        High evaporation rates

Problem for farmers:
        Lack of water to grow crops
        Soil becomes salty
        Isolated areas

Solutions:
        Build wells and pump water for irrigation

Problem:
        Costly
        Low yields
        Little grass for grazing
        Food shortage

Population density:
        Few people are attracted to
        live in hot deserts
        Hot deserts can only support a
        low population density

Figure 2.4 Why hot deserts have a sparse population

TASKS

1 Using an atlas, locate Areas 2–6 shown on Figure 2.5. For each area decide whether it would be a high or low density and suggest why. Area 1 is done for you.

Location	High or sparse density	Reason
Andes	Low density	Mountainous. Too steep and cold for farming.

2 Explain why there is a link between environmental factors and population density.

3 Design a flow chart to show:
   a why deltas attract high population densities,
   b why cold deserts have a low population density.

4 Explain how occupation structure can affect the population distribution.

5 On an outline map of the world, shade the areas that are most suitable for people to live. Use an atlas to find additional information, i.e. climate and relief. Write a description of your map and explain how you decided on your areas.

FIGURE 2.5
Natural environments

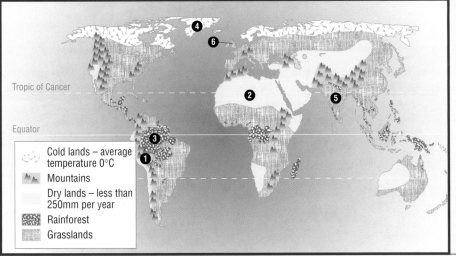

Tropic of Cancer

Equator

Cold lands – average temperature 0°C
Mountains
Dry lands – less than 250mm per year
Rainforest
Grasslands

POPULATION DISTRIBUTION IN THE UK

Figure 2.6 shows the population distribution for the UK. This map provides more detail about the variation between different parts of the UK than Figure 2.2. The UK now has a population of 58 million people and an average population density of 243 per sq. km.

In England, the high population density is due to urbanisation. 89% live in urban areas. At least half of these people live in **conurbations**. Figure 2.6 shows an axial belt of dense population extending north-west from south-east England, through the West Midlands to Manchester and Liverpool in the north-west and eastwards to Bradford and Leeds. Major rail and road links connect all these areas. Further north, there are the conurbations of Tyneside and Central Scotland. There are also large urban areas in South Wales, including Cardiff and Swansea. All developed as major industrial centres, often located on or close to coalfields that provided power for their industries or with ports to import raw materials and export goods.

These high density areas are surrounded by rural landscapes. These include the highland and less accessible areas. Many lack industrial raw materials and are agricultural.

Country	Area (sq. km)	Population	Pop. density
England	130 423	47 million	376
Wales	20 766	2.8 million	141
Scotland	78 133	5 million	66
N. Ireland	14 160	1.5 million	117

Figure 2.8 Population data for the UK in 1991

FIGURE 2.6
Population density

FIGURE 2.7
Highland areas and major settlements

Derbyshire's population pattern

Population distribution in Derbyshire has been influenced by the **relief** of the Peak District and economic activity based on the coalfields in the north-east and south-east of the county. The Peak District forms a sparsely populated area of limestone hills in the west where the vegetation is rough grazing and moorland. The higher land is cooler and wetter than in the east of the county. Industrial towns like Chesterfield and Alfreton developed on the coalfield and better agricultural land. The main railway lines, main roads and now the M1, help to create a continuous line of settlements on the eastern side (Figure 2.9).

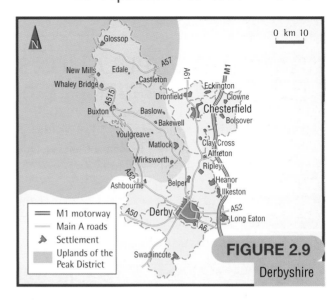

FIGURE 2.9 Derbyshire

TASKS

1 Describe the distribution of large urban areas in the UK. Suggest two factors which have influenced this distribution.

2 Present, using an appropriate method, the information in Figure 2.8. Describe what your graphs show.

3 What is the best way to show population distribution? Why is this the best?

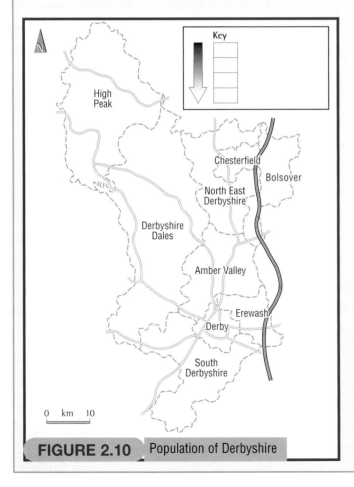

FIGURE 2.10 Population of Derbyshire

4 Make two copies of Figure 2.10. Choose colours or shading to show 'the higher, the darker'. Add a key to each map.

a plot the actual population data using this scale,
   above 200 000
   100 001 to 200 000
   80 000 to 100 000
   less than 80 000.

b plot the density of population data using this scale:
   above 10 per sq. km
   5.1 to 10 per sq. km
   3 to 5 per sq. km
   below 3 per sq. km.

c Describe and explain the pattern shown by each map.

District	Population 1991	Population density numbers per ha.
Amber Valley	111.9	4.1
Bolsover	70.4	4.3
Chesterfield	99.4	15.1
Derby	218.8	27.4
Derbyshire Dales	67.6	0.9
Erewash	106.1	9.5
High Peak	85.1	1.5
N E Derbyshire	97.5	3.5
S Derbyshire	71.8	2.1
County (average)	28.6	3.5

Figure 2.11 Population data for Derbyshire

5 Choose an area of high population density in the UK. Use the thematic maps in an atlas and other sources to explain reasons for this dense concentration of people.

6 Locate and identify an area of sparse population in the UK. Explain how its location, relief, climate, raw materials and economic activity have affected its population density.

CHANGING POPULATIONS

By 2050, the world's population is predicted to be almost 9 billion. In AD 1 the population was 0.3 billion, by 1800 it was still less than 1 billion. It took 127 years to double to 2 billion. However, it only took 46 years to double to 4 billion (Figure 2.12). This is a rapid growth rate which now adds 78 million people to the world's population each year. Why is world population rising so rapidly? What population can the earth sustain?

Year	Population (billions)
AD 1	0.3
1000	0.4
1600	0.5
1700	0.7
1800	0.9
1900	1.7
1950	2.3
1960	3.0
1970	3.9
1987	5.0
1999	6.0
2010	8.0
2050	8.9

Figure 2.12 World population growth

The rate of population growth varies between regions and countries. The highest rates are found in LEDCs (Figure 2.13). Change is the result of **migration** and **natural increase**.

Natural increase = birth rate

death rate

Continent	Birth rate	Death rate	Natural increase	Life expectancy	Doubling time
Asia	24	8	1.6	65	44
Africa	40	14	2.6	53	26
North America	14	9	0.6	73	117
South America	25	7	1.8	69	38
Europe	10	12	-0.1	70	–
Oceania	19	8	1.1	74	63

Figure 2.13 Population growth rates

This process of change has four stages. They are shown in the **Demographic Transition Model** (Figure 2.14). It suggests that over a period of time birth and death rates fall as medical care, living conditions and education levels, particularly of women, have improved. In some countries family planning has reduced birth rates. People are also living longer as their quality of life improves.

MEDCs have reduced their natural increase to almost zero, and several European populations are in decline. Japan's population is expected to halve in the next 50 years.

FIGURE 2.14

Demographic Transition Model

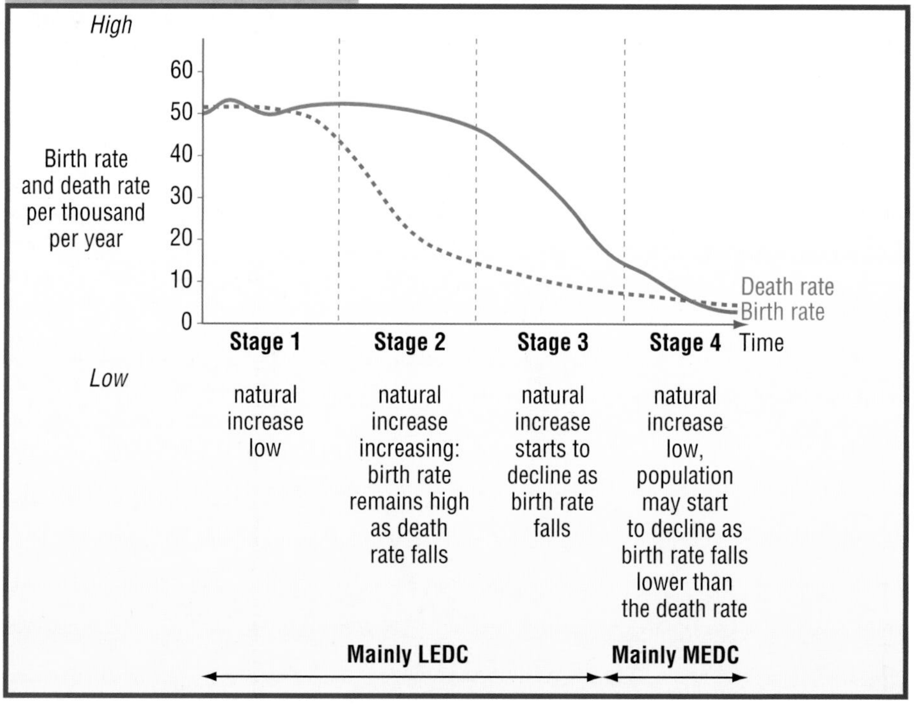

LEDCs have high rates of natural increase, for example, Nepal 2.3, Uganda 2.9 and Bolivia 2.8. An increase of 2.0 will mean the population will double in just 34 years.

MEDCs have moved into Stage 4 of the transition model. LEDCs tend to be found in Stages 1 or 2. Their lack of economic development restricts the educational and social improvements needed to reduce birth rate and family size. Whereas in MEDCs the average **fertility rate** is 1.6 children per woman, in LEDCs it is 4. In some countries, for example Niger or Uganda, it is over 7. The world average is 3.

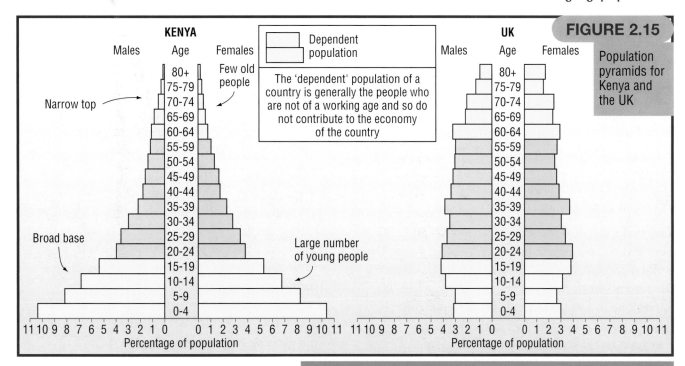

FIGURE 2.15

Population pyramids for Kenya and the UK

Countries at different stages of the demographic transition model have different population structures. Population pyramids present this information graphically. The population structure identifies the number of males and females in different age groups and shows those who are **dependant**, usually too young or old to work. To find out the **dependency ratio**:

Dependency ratio =

(number of children under 15 + number of people over 65) × 100

Number of those of a working age 16–64

In MEDCs, the dependancy ratio is usually between 50 and 70 as families tend to be smaller. However in LEDCs the ratio often exceeds 100 because families include large numbers of children.

This is shown in the population pyramid for Kenya (Figure 2.15). It has a narrow top and broad base. The narrow top is because the death rate is high with few people surviving to old age. The high birth rate means that there will be large families. By contrast the UK has a narrower base and wider top. This reflects the low birth and death rate, and that people live longer.

TASKS

1 Use Figure 2.12 to draw a line graph to show world population growth. Annotate the finished graph to highlight: slow growth, rapid growth, long doubling time, short doubling time.

2 What is the rate of natural increase for these countries:

Country	Birth rate	Death rate
Nigeria	43	13
Mali	50	20
Germany	10	11
Italy	9	9

3 Explain why Italy and Germany have different rates compared to the others.

4 For each of the four stages of the demographic transition model, describe how the birth rate and death rate changes.

5 Consider the demographic transition model for Kenya and the UK. At which stages are Kenya and the UK? How might they change in the next 50 years?

6 Compare Kenya's population structure to that of the UK. You should consider:

    a   the proportions in the different age groups,
    b   the dependency ratio,
    c   reasons for the differences.

7 Explain why a country can still have low birth and death rates yet still have an expanding population.

8 The US census website includes a section which allows you to select population pyramids for any country for different years. http://www.census.gov/ipc/www/idbpyr.html

    a   Use the website to select pyramids for an MEDC and an LEDC.
    b   Copy and paste these pyramids into a desktop publishing program.
    c   Use the software tools to label the characteristics of each pyramid.
    d   Add which stage of the demographic transition model the country appears to be in.

POPULATION ISSUES

The rapid growth of the world's population has raised many concerns and questions. Is it a problem of too many people, or not enough resources? Some experts are concerned that the ecosystems that help keep the earth healthy are breaking down as rapid population growth, exploitation of resources and pollution affect the environment.

Every minute, 170 babies are born, while only 80 people die. This means by the end of each year, 78 million more people will be needing food, water, shelter and work. Most of these will be in LEDCs, where one in five go hungry. The 25% of the world's population who live in the MEDCs consume about half of the world's food and use 80% of the energy.

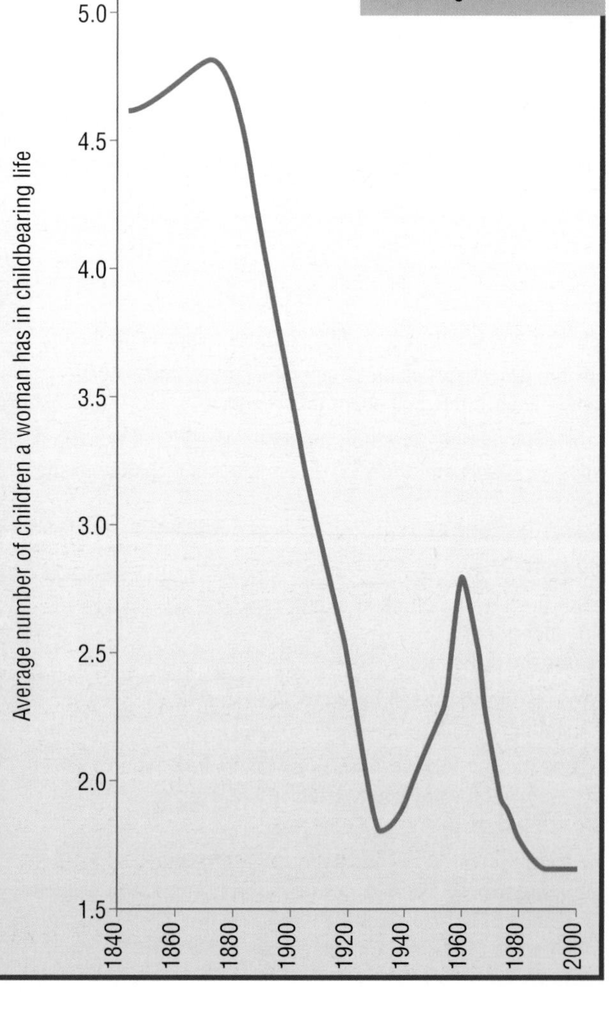

Most MEDCs have low natural increase. Many fall below the 2.4 fertility rate needed to keep the population number stable. A combination of family planning, education, career ambition and a high standard of living has seen family size fall (Figure 2.16). In the next 40 years the average age of people in the UK will rise from 38 to 42 years, the dependency ratio will increase from 58 to 70 and there will more than 3 million people over 80 years old. It is predicted that in 2008 there will be more pensioners living in the UK than children. There will be 2 million fewer in the 16–60 age range. This means fewer workers and less taxes for government to spend providing social and health services to care for the UK's ageing population. This **greying population** is a feature of MEDCs (Figure 2.17).

FIGURE 2.16

Shrinking UK families

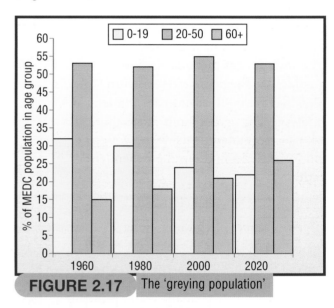

FIGURE 2.17   The 'greying population'

LEDCs are changing too (Figures 2.18 and 2.19). The average age there will rise from 26 to 37, the number of elderly will double, particularly women who live longer than men, and governments will struggle to provide higher standards of care. LEDCs' populations will become more urban. While there will be a lower dependency ratio, the lack of workers in MEDCs will encourage migration from LEDCs to fill the gaps in the economy. It is expected that LEDCs' population structures will be more stable but will still continue to grow. Many LEDCs have natural increases above 3 per thousand per year.

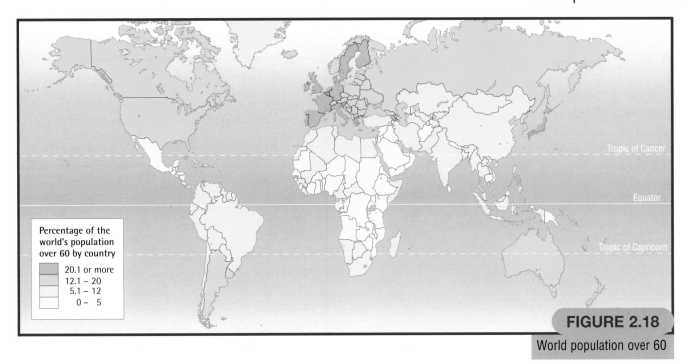

FIGURE 2.18

World population over 60

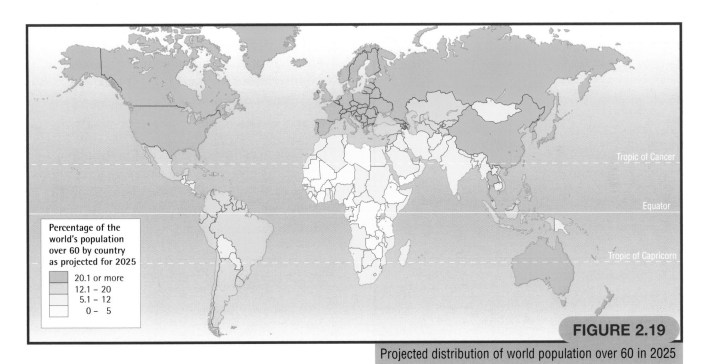

FIGURE 2.19

Projected distribution of world population over 60 in 2025

Birth control

In Africa, 60% of all births are to mothers below 20 years old. To reduce their high natural increase, many LEDCs have encouraged family planning schemes. Some schemes have been forced on their populations, for example, in China. Others rely on educating women, for example, Mexico or Thailand. By encouraging women to make educated decisions about the size of the families and providing medical support for safe motherhood and ensuring children survive, several countries are now reducing their growth rates. Surveys suggest that the more educated a woman, the fewer children she is likely to have and the healthier they will be. The 1994 Cairo Conference proposed that improving the status of women was the key to reducing birth rates. Cultural and religious barriers can make this more difficult.

Population and the environment

The gap between the richer and poorer nations is growing. But as the LEDCs try to catch up with the MEDCs, there will be a considerable impact on the environment. Many areas are **overpopulated** and the effort to make a living is damaging the environment, for example, desertification in the Sahel, in Africa. Despite having a sparse population there are too many people for the available food, water or energy resources to support. At least 50 countries are described as having a water scarcity. This means they cannot provide the annual 1 700 cu. m of water that each person needs to survive. The lack of water affects health and the ability to irrigate food crops.

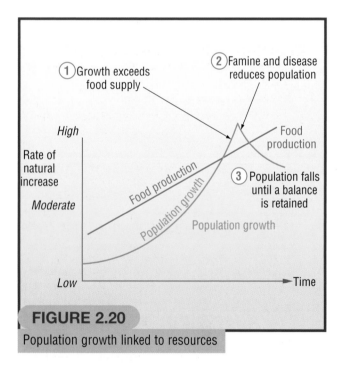

FIGURE 2.20

Population growth linked to resources

The link between food supply and population can be graphed (Figure 2.20). It suggests a steady increase in the amount of food and if population exceeded this amount, there would be disease and famine which would reduce the numbers. However, the evidence suggests that there is sufficient food to provide everyone with an adequate diet despite the increases. But there is 10% less food available in Africa now than ten years ago. Many people will be hungry all their lives. Wars only make things worse. Population experts predict two possible outcomes (Figure 2.21).

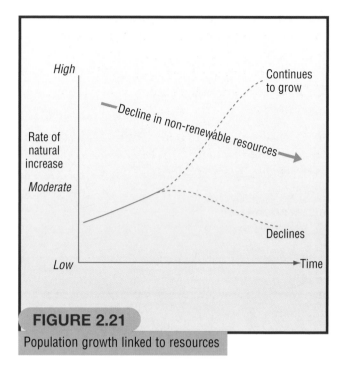

FIGURE 2.21

Population growth linked to resources

▶ The population will stabilise or decline so that population and resources become balanced, or

▶ the population will continue to grow until it exceeds the environment's capacity to carry it.

The result would be rapid decline in the population as the earth's non-renewable resources are used up and the economic base collapses. Food shortages and environmental damage would lead to a rapid decline in the population.

TASKS

1 Describe how populations in MEDCs may change in the next 50 years.

2 List three social and economic consequences of these changes.

3 Which change do you think will have the greatest impact on the UK? How and why?

4 Why did the Cairo Conference recognise the importance of women in changing attitudes towards family planning? Suggest some ways people are able to reduce their family size.

5 What is the link between environmental degradation and overpopulation?

6 Use Figure 2.21 to predict what you think is the most likely outcome of world population growth. Suggest how you think this issue should be tackled.

FAMILY PLANNING IN CHINA

At one time, China encouraged large families to increase the country's workforce. It was soon realised, however, that large population increases could not be sustained. Since the 1960s, the Chinese government has enforced laws about family planning to reduce the growth rate.

In 1975, the average Chinese family had three children. Chinese officials predicted that, left unchecked, China's population would rise to 2 billion in the next hundred years. In 1979, it was decided that a 'one child per family' policy should be introduced.

The policy was forcibly implemented and a culture of anti-social behaviour was attached to parents who had more than one child. They could lose benefits and be fined. In many cases, people were made to have abortions and be sterilised. The government provided incentives such as child care to reward one-child families. There was constant propaganda about the benefits of having only one child.

The policy had its greatest impact in the cities. By 1986, 83% of families in the cities had only one child. In the countryside, however, the policy has been less successful where only 62% of families have one child. Overall the fertility rate has been reduced to 1.9 (Figure 2.22).

Date	Birth rate	Death rate	Pop. change	Population (m)
1950	44	25	1.9	554 760
1960	38	17	2.1	657 492
1970	31	9	2.2	830 675
1980	19	7	1.2	996 134
1990	20	8	1.2	1 155 305
2000	17	7	1.0	1 300 000

Figure 2.22 Population change in China

In rural areas there is a tradition of having large families to help with farming and support parents in their old age. Children provide free labour. As the rural areas become more prosperous, people were prepared to pay the fines for having a large family, particularly if it meant they had boys.

The government compromised and allowed rural families to have a second child if the first was a girl.

This meant that by 1986 China had 14 million more children than originally planned for in 1979.

While the policy has reduced China's growth, it has raised several other issues:

▶ China will have an age imbalance and not have enough workers in the future.
▶ China's population could decline.
▶ Girl babies were undervalued and have been killed so that their parents could try for a boy.
▶ The strict rules caused unrest.
▶ The one child, two parents and four grandparents family was becoming common.

TASKS

There are a number of Internet sites which explore world population issues.
- 6 Billion Human Beings, an Interactive Exhibition, produced by Muséum National d'Histoire Naturelle Paris – France:
  http://www.popexpo.net/english.html
- The World Overpopulation Awareness website is a forum for people to share ideas about population issues
  http://www.overpopulation.org
- UNFPA, the United Nations Population Fund website, includes an interactive population centre which gives a wide range of information about world population issues:
  http://web.unfpa.org/modules/intercenter/index.htm

1 Refer to the above websites to research world population issues, and to produce your own report outlining the issues and suggestions for solutions, using a word processor.

The State Family Planning Commission of China has produced a website, promoting their family planning policies; use this to help you complete the tasks below.
http://www.sfpc.gov.cn/epopindex.html

2 Suggest why family planning policies were introduced in China.
3 Describe how China implemented its family planning policy.
4 List the successes and problems of China's policy.
5 Present the information in Figure 2.22 as appropriate graphs or charts. Describe the impact of family planning on the population since 1950.
6 Why do you think that family planning policies are needed in many LEDCs? What are the alternatives?

MIGRATION

International migration is the movement of people from one country to another. **Emigrants** are people who leave a country to settle in another. **Immigrants** are the people who arrive to live in another country. It is estimated that in the last 40 years over 35 million people have moved from LEDCs to MEDCs to live. Migration has been made easier and more available with improvements in international transport.

People who emigrate do so for a variety of reasons. Many move to find a better quality of life and employment. There is an incentive for the poorest 20% who earn only 2% of the world's GDP to move to the MEDCs where over 80% of the world's GDP is produced. Upwards of 6 million illegal immigrants live in MEDCs. Many MEDCs have a shortage of workers and immigrants help fill many of the low-paid manual jobs. Others move for family reasons, or as a result of a natural disaster or conflict. Movement can be voluntary, for example, in Germany guestworkers make up 7% of Germany's population, or because of colonial links for example, West Indians who moved to the UK, or forced as refugees to leave, for example, Albanians from Kosovo.

Guestworkers in Germany

Post-war Germany received 13 million people from Eastern Europe. This stopped when the Berlin Wall was built in 1961. However, Germany still attracted immigrants to work from the poorer Southern European countries (Figure 2.23). The immigrants could expect to earn a lot of money, some of which was sent home to support their families and invest in businesses. These immigrants were regarded as **guestworkers** as they were expected to return home, but many decided to settle in Germany and give their families a better quality of life.

UK	110 000
USA	110 000
Poland	230 000
Italy	600 000
Turkey	2 000 000
Yugoslavia	800 000
Greece	360 000
Portugal	125 000
Spain	130 000

Figure 2.23 Origin of Germany's guestworkers

Immigration in the UK

During the 1950s the UK had a labour shortage, while unemployment was high in the West Indies. Many West Indians had sought work in the USA, but restrictions imposed by the American government limited immigration. Many West Indians held British passports and these allowed members of the Commonwealth to enter the UK. Several organisations, such as London Transport and the National Health Service, set up offices in the Caribbean to recruit workers. During the 1950s and 60s, there was also an influx of immigrants from other Commonwealth countries which were also LEDCs, for example, India (Figures 2.24 and 2.25).

Country	1990
European Union	66 000
Australia/New Zealand/Canada	57 000
South Asia (India, Pakistan)	22 000
Caribbean	7 000
USA	29 000
South Africa	6 000
Middle East	10 000
Others	70 000

Figure 2.24 Migration into the UK

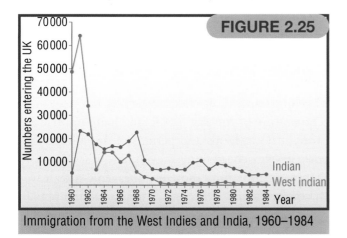

Immigration from the West Indies and India, 1960–1984

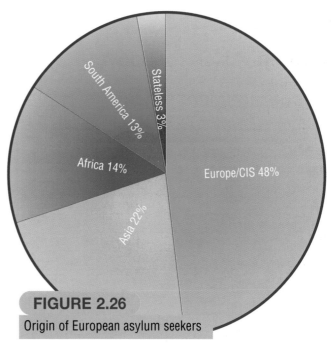

FIGURE 2.26

Origin of European asylum seekers

The arrival of immigrants and their families caused concern to some people and increased racial unrest. Some people thought immigrants were taking jobs away and putting pressure on the welfare system. Many faced discrimination and hardship. The British government came under pressure to restrict the numbers entering the country. New laws used family rather than economic criteria to allow people to settle in the UK. Laws were passed to protect and support immigrants to become successful.

There are 26 million refugees worldwide. Until a person is granted refugee status by a country he or she is known as an asylum seeker. A **refugee** is a person who has left their country through fear of persecution. Most refugees move to neighbouring countries, for example, ethnic Albanians move from Kosovo to Albania, Greece, Bosnia and Macedonia. From there, some were airlifted to Germany, UK and other European countries. The effect of unrest in a region is often made worse by poverty and environmental issues (Figure 2.26). In the Sahel, in Africa, drought and desertification has caused thousands of people to leave. This has created a new group of economic and environmental refugees who move to find better living conditions.

There is also migration from MEDC to MEDC, for example, from the UK to Australia (Figure 2.27). This has much to do with a quality of life and desire to have a new start somewhere not too different from the home country.

Country	1994
UK	10 300
New Zealand	10 200
China	7 200
India	4 900
SE Asia	10 500
Yugoslavia	3 000
CIS	2 300
USA	4 500
Africa	2 300
North Europe	1 400
South Europe	520
Others	23 400

Figure 2.27 Origin of immigrants to Australia, 1994

TASKS

6 Make a list of reasons why people decide to emigrate. Which reasons are positive reasons for emigrating?

7 If you had a choice to emigrate, where would you go, why and what difficulties would you have to overcome?

8 Explain why so many immigrants from LEDCs choose to live in an MEDC.

9 What is the difference between refugees and voluntary immigrants?

10 Plot Figure 2.24 as a flow line map to show the origin of immigrants in the UK. Compare this with Figure 2.27. Explain the differences.

11 Why was the UK attractive to West Indian immigrants?

12 Do you agree that MEDCs should accept immigrants from LEDCs? What are the benefits to the MEDC? What problems might the immigrants have to deal with? How might the LEDC be disadvantaged by the loss of people?

MEXICANS TO USA

Encouragement to leave

Mexico and USA are next to each other on the map, but they are worlds apart. Look at the differences in Figure 2.28.

	USA	Mexico
Births per 1000 population	15	27
Years for population to double	116	32
Infant mortality	7	34
Percentage of population under age 15	22	36
Percentage using modern contraception	68	56
GNP per person in dollars	26 980	3 320

Figure 2.28

Mexico is an LEDC and USA is an MEDC. The birth rate in Mexico is almost double that of the USA. More than one third of Mexicans are under 15 years old and yet to have families, so the population is likely to double in only 32 years. Already there are problems of providing everyone with enough food. Farmers are now trying to grow crops on land they used to consider as too poor for agriculture. Increasingly, there is not enough land to farm. It is a **push factor**, encouraging people to leave.

El Niño is a weather effect caused by the temperature of the Pacific Ocean changing. It has brought both floods and periods of drought to parts of Mexico, making it even harder to earn a living and providing more encouragement to leave. In country areas, living conditions are often poor. 80% of people have no clean water supply and about half of the children in rural areas have left school by the age of eleven, most unable to read and write.

The attraction of the USA

From a very early age, Mexicans have a glamorous idea of life in USA – the 'Coca-Cola' image. Los Angeles, just 250 km across the border, is known as the 'City of Opportunity'. The wealth and bright lights image of USA are **pull factors**. Fifty years ago, almost all migration from Mexico to USA was seasonal. A few of the men from the villages near the border would cross to pick the cotton crop in Texas, fruit and vegetables in California's Central Valley or work in the food processing factories. They would return to their families a few months later with as much money as they could earn in a whole year in Mexico. Today, there still is seasonal work and USA could not manage without Mexican

FIGURE 2.29

Mexicans provide cheap labour for low skill jobs in California migrants. Work permits are issued for the number of workers needed. These are the legal immigrants. There are many more Mexicans who want to get into USA.

No one knows how many illegal immigrants cross the border each year. It is estimated there could be 2 million attempts, with 300 000 people managing to get past the Border Patrol agents. Some parts of the border have a 4-m-high metal fence and searchlights. Although there are 5 500 Border Patrol agents, the border is 3 326 km long. Even with tracker dogs and helicopters, it is an impossible job. Those caught are deported back to Mexico – and try again next night.

FIGURE 2.30

Maquiladora factories pollute and encourage migration

Most migrants head for places where they have contacts, such as relatives or someone from the same village. This will help them find work and accommodation. Almost half of all migrants go to California. They live in the poorest housing, in communities almost entirely of Mexicans. Many Californians resent them. They fear that, as communities become established, Mexicans will work their way up to take better jobs, as Asian immigrants have done. At present, Mexicans are the 'working poor'. They work long hours for low pay by American standards, doing the dirty, dangerous, boring or unpleasant jobs that are hard to fill.

Maquiladora developments

In the last 20 years, more than 2 000 new factories have been built just inside Mexico for American firms. They are called **maquiladora** developments. The attractions are:

▶ Mexicans are willing to work for wages much lower than would have to be paid in USA.

▶ Production costs are lower because there are fewer environmental controls in Mexico so waste can be dumped cheaply.

▶ Being close to the border, goods can easily be taken to USA for sale.

These new factories, however, have encouraged migration from the Mexican countryside and provided some experience of working in industry.

These people then become more likely to make the short journey across the border into USA, in search of a better life.

Mexico's loss

It is not the poorest and most destitute who migrate. Most can read and write. They have some skill that will make them employable. A migrant may have worked previously in Mexico as a bricklayer or a driver, for example. So the villages lose their most trained and best educated, as the poorest and least adaptable remain in the countryside.

TASKS

1 Study the information in Figure 2.28.

   a At the present rate of increase, in which year will the population of USA be double what it is now?

   b In which year will Mexico's population be double what it is now?

   c Use all the information in the table to suggest reasons for Mexico's population increase.

2 In two columns, list the *Push Factors* in Mexico and the *Pull Factors* of USA.

3 What is the attraction to Mexicans of California? Begin with its location.

4 What are the advantages to a new migrant of heading for a community of other migrants from Mexico?

5 Migration does not appeal to all equally. What suggests that pull factors are stronger than push factors?

People and Places to Live

Settlement

WHAT IS URBANISATION?

Urbanisation is the process by which the number of people living in cities increases compared with the number of people living in rural areas. A country is considered to be urbanised when over 50% of its population lives in towns or cities. In 1800 only 3% of the world's population lived in urban areas, but by 1950 this had risen to 29%. It is estimated that this proportion may rise to over 50% by 2006. Figure 2.31 shows the proportion of each country's population living in urban areas.

Among the first countries to become urbanised were the UK and some other European countries.

Urbanisation here was relatively slow, allowing governments time to plan and provide for the needs of increasing urban populations. Today, urbanisation is most rapid in LEDCs. It took London's population 100 years to expand from 1 million in 1860 to 9 million in 1960. By contrast, Mexico City's population was 1 million just 50 years ago, and is over 15 million today. In Asia, the urban population is expected to reach 2.5 billion in 2025, three times what it was in 1990. This growing shift in the world distribution of urban populations is shown in Figure 2.32.

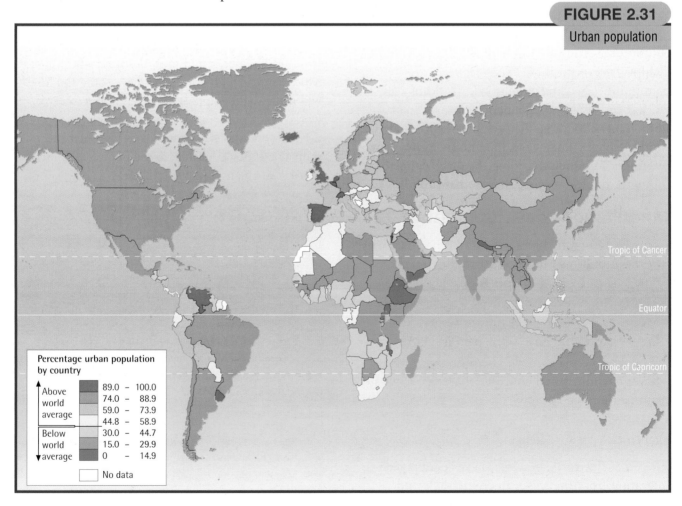

FIGURE 2.31

Urban population

Percentage urban population by country

Above world average
- 89.0 – 100.0
- 74.0 – 88.9
- 59.0 – 73.9
- 44.8 – 58.9

Below world average
- 30.0 – 44.7
- 15.0 – 29.9
- 0 – 14.9

No data

Tropic of Cancer

Equator

Tropic of Capricorn

Megacities

A recent feature of urbanisation has been the growth of really large cities. In 1900, there were only two cities with a population of over 1 million (millionaire cities), London and Paris. Today there are over 300 millionaire cities. The ten largest cities in the world have populations over 10 million – these are called megacities. By 2015 there will be 27 megacities, 22 of them will be in LEDCs. A total of 87 Asian cities, including 38 in China and 23 in India, have more than 1 million inhabitants. Latin America is even more urbanised than Asia, while Africa is slightly less so. Across the world, not only is the number of large cities increasing, but smaller cities are becoming larger at a faster rate.

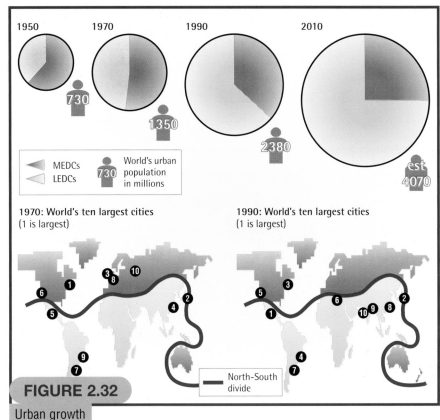

FIGURE 2.32
Urban growth

TASKS

1 Look carefully at the map of the world's urban population (Figure 2.31).
  a Which continents have the highest proportion of urban population?
  b Which continents have the lowest proportion of urban population?

2 Look carefully at the resources in Figure 2.32.
  a Describe what is likely to happen to the world's total urban population between 1950 and 2010.
  b Draw two line graphs on the same axis to show the numbers of people living in urban areas in LEDCs and MEDCs between 1950 and 2010. Plot the year on the x axis and the numbers of people living in urban areas on the y-axis. Use different colours for each line and show what these colours represent on a key.
  c Describe any changes that you can identify.
  d How did the top ten largest cities in the world change between 1970 and 1990?

3 a Figure 2.33 shows a United Nations estimate for the largest 15 world cities for 2015.
  b Mark the locations of these cities on an outline map of the world.
  c Add the North–South divide to your map.
  d Add the names and projected populations of each city, together with its rank position.
  e Describe and explain any changes to the distribution of the world's largest cities since 1990.

Figure 2.33 Projected 15 largest cities in the world, 2015

	(pop. in millions)
1 Tokyo, Japan	28.7
2 Bombay, India	27.4
3 Lagos, Nigeria	24.4
4 Shanghai, China	23.4
5 Jakarta, Indonesia	21.2
6 São Paulo, Brazil	20.8
7 Karachi, Pakistan	20.6
8 Beijing, China	19.4
9 Dhaka, Bangladesh	19.0
10 Mexico City, Mexico	18.8
11 New York, USA	17.6
12 Calcutta, India	17.6
13 Delhi, India	17.6
14 Tianjin, China	14.7
15 Metro Manila, Philippines	14.7

Source: United Nations 1995

RURAL TO URBAN MIGRATION IN LEDCs

Pull

Improved employment opportunities, jobs in factories pay more than farming

Improved quality of life and standard of living to migrants who not only want a better life for themselves but are determined that their children should have the chance of a good education to benefit from the advantages this might bring them.

Expectations of improved housing with services such as electricity and water

Availability of schools, hospitals, and entertainment.

Access to education is very important

FIGURE 2.34

Push and pull factors

Push

Pressure on the land not enough for people to live on

Large families mean that there is not enough land for each child

Poor quality of life – hard work, long hours, little pay for farmers

Poor-quality housing

Increasing use of machinery reduces job opportunities in farming

Lack of infrastructure eg. electricity, water and sewage services

Overpopulation due to high birth rates

Starvation

Natural disasters e.g. drought, hurricanes, floods and volcanic eruptions, crop failure, fire, pests

Workers do not own the land and feel powerless

Lack of education, health and welfare facilities

Lack of investment from the government

The growth of cities in LEDCs is partly due to the high rate of natural increase in population. This is compounded, however, by a high rate of migration from rural areas. It is estimated that 3 000 migrants arrive in urban areas of LEDCs every hour of the day. Some of the reasons for moving from the countryside are shown in Figure 2.34. Poverty and lack of opportunities push people from the countryside to the cities. There are two groups of factors – **push** and **pull**, the decision to move is usually a result of the interaction of factors from each group. For example, a farmer in rural Brazil whose land is suffering from drought and is increasingly

unproductive, would not be 'pushed' off his land and decide to move to São Paulo unless he was also aware of opportunities to improve his income there (pull).

People are lured to the cities because they believe they will have a better quality of life with job opportunities, and a future for their families with better access to education and health care. Improved telecommunications – radio, film, television – and better transport have made the rural population more aware of the potential opportunities in cities. This has accelerated the process of urbanisation in LEDCs.

Direction	Migrants per 1000 of the population	
	1995	1996
Rural to rural	11.32	10.72
Rural to urban	7.8	8.3
Urban to rural	0.87	0.83
Urban to urban	21.29	31.36

Figure 2.35 Movement of migrants in Bangladesh

	Rural	Urban
Infant mortality rate	76/1000	50/1000
Life expectancy (age)	58.2	61.2
Adult literacy rate (%)	38.6	64.4
Calorie intake (kcal)	2283	2240
Access to toilet (%)	6.2	32.1
Access to safe drinking water (%)	95.8	99.4

Figure 2.36 Urban and rural contrasts in the quality of life in Bangladesh, 1996

Some migrants move direct to cities from the countryside, others move to larger villages and towns, this is called **step migration**. Many migrants move to join relatives and friends who are already established in the city. Their perceptions of what the cities are like often reach them second- or third-hand. Anyone thinking of migrating any great distance is unlikely to have enough money to go and visit the city beforehand to see for themselves what it is like. Often their ideas of what city life will be like are unrealistic. The people who move tend to be the most educated and ambitious in the rural community. They have the drive necessary to take this bold and life changing step. These are the very types of people that the rural communities need to overcome their problems. Their departure heralds a spiral of decline for rural areas.

Bangladesh Official Government site:

http://www.bangladeshonline.com

includes a data sheet with a wide range of statistics for Bangladesh.

PROBLEMS OF CITIES IN LEDCs

Once the migrants arrive in the cities, they find life very different to what they imagined. The rapid influx of people, most of them extremely poor, has caused many problems for cities in LEDCs, in particular there is an acute shortage of housing. Vast areas of squatter housing have grown up. In Brazil, they are called favelas; in India, they are called bustees; in Africa, shanties. These settlements are usually illegally built up almost overnight on any vacant plot of land near the edge of the city, along rail or road routes, or on vacant land close to the city centre, often on unsuitable sites near factories on steep slopes, near waste sites or swamps. The areas used are often avoided by other land uses because they are prone to landslips, flooding or industrial pollution.

The houses are home-made, built from anything that people can find, including waste wood, sheets of corrugated metal, cardboard, polythene and oil drums. They represent a major fire hazard. These are unplanned, haphazard developments. As soon as one family squats on a piece of vacant land, many more follow. The conditions are cramped and crowded. Their homes are typically one-roomed dwellings in which the whole family has to live, eat and sleep. Some people have no home at all, sleeping in shop doorways, alleys or under railway bridges.

TASKS

1 Look carefully at Figure 2.35.
  a Which type of migration seems to be most important?
  b Explain the reasons for this pattern.

2 Read the case study of flooding in Bangladesh on pages 36–39. Additional information can also be obtained from the Internet. The Bangladesh Government has created a site about the 1998 flood:

  http://www.bangladeshonline.com/gob/flood98/

  Explain why successive floods might act as a push factor encouraging people to leave the countryside.

3 Use the statistics in Figure 2.36 to write a paragraph about the quality of life in rural and urban areas in Bangladesh.

4 The photographs on the next page show typical images in rural and urban Bangladesh. Use the photographs to create a list of push and pull factors for Bangladesh. What push and pull factors can you identify from the photographs?

5 In what ways is migration within an LEDC different from the migration of Mexicans to USA? (pages 80–81)

Lack of basic amenities

The government and city authorities cannot keep up with the pace of urban growth. Most homes in the squatter communities lack basic amenities such as electricity, gas, running water and a sewage system. In the bustees of Calcutta, it is common for one water tap and toilet to be shared by 30 people. Raw sewage runs down the streets, polluting the water supply. Diseases such as cholera and typhoid spread quickly in the hot, cramped, unhygienic conditions prevalent in the squatter communities of LEDCs. Every year, 4 million urban residents worldwide die from water-borne diseases alone. In Manila, the capital of the Philippines, one researcher found 35 diseases that were caused by garbage and filth (filth-borne diseases), five of which were the top five killer illnesses in the country. There is often no refuse collection and any spare space quickly becomes filled with rubbish, which is another breeding ground for disease.

There is often a lack of employment opportunities in the squatter settlements. The new arrivals are often poorly educated with few skills. Many join the informal sector of employment selling vegetables on the streets or collecting rubbish for recycling.

Living in shanty towns is very stressful. Many marriages break down as a result and crime, especially vandalism and theft, is rife. Many children are abandoned by their parents and live on the streets.

Many of the world's fastest growing cities are expanding into areas unsuitable for large-scale human habitation, susceptible to natural disasters such as flooding, landslips and earthquakes, making the risk of heavy loss of life a reality. The BBC News reports from the Internet highlight this threat.

BBC News

Friday, August 20, 1999 Published at 17:13 GMT 18:13 UK

Deathtrap cities keep growing

By Environment Correspondent Alex Kirby

One of the causes of the enormous loss of life in disasters in our modern age is the growth of modern cities. One hundred years ago, London headed the list of the world's ten largest metropolitan areas with 6.5 million inhabitants. But the UK capital does not even appear on the list of the ten largest cities on earth in the year 2000. Most of the world's biggest cities are in countries far poorer than the UK – places like Bombay, São Paulo and Lagos. Many cities are poor at the best of times. They are so jam-packed that, for millions of their people, day-to-day life is already lived wretchedly on the margins. A recent report from the Worldwatch Institute, *Rein enting Cities for People and Planet*, gave an outline. 'At least 220 million people in cities of the developing world lack clean drinking water, and 420 million do not have access to the simplest latrines. Six hundred million do not have adequate shelter, and 1.1 billion choke on unhealthy air.'

When chronic poverty erupts in natural or man-made disaster – earthquake, tidal wave, fire, storm or any other horror – the numbers of people in modern cities force casualties off the scale. The pace of urbanisation means conditions in the cities are getting worse much more quickly than governments can act. Between 1990 and 1995, Worldwatch says, 'the cities of the developing world grew by 263 million people – the equivalent of another Los Angeles or Shanghai forming every three months'. With almost half the people in the world living in cities today, the problem is already upon us.

In an ideal world, buildings would be constructed to survive earthquakes. The technology is there, but the money and the will are not. Probably the best we can hope for is not protection, but more preparedness. Next year marks the end of the UN's international decade for natural disaster reduction.

It exhorts all countries by 2000 to have comprehensive national assessments of risks from natural hazards integrated into development plans to address long-term disaster prevention, preparedness and awareness ready access to global, regional, national and local warning systems.

Perhaps the simplest warning system might be to advise intending urban migrants they would often be safer staying put in the countryside.

BBC News

Thursday, July 1, 1999 Published at 09:42 GMT 10:42 UK

Bangladesh struggles to cope

Overcrowded cities as many Bangladeshis leave their villages

By David Chazan in Dhaka

Zaynab Begum, her husband and three children live in a tiny bamboo and corrugated hut in a Dhaka slum. Their living space is smaller than six square metres. They have no electricity, running water or toilet, and an open sewer runs outside the hut. 'It's difficult to live here because it's cramped and uncomfortable,' she said. 'But my husband is a rickshaw-puller. We can't afford anything else.'

Zaynab Begum's family are among millions of Bangladeshis who have migrated to the cities in search of work. Fifty years ago, only about 4% of the population lived in urban areas. Now more than 25% are city dwellers. Overcrowding and malnutrition are worst among the urban poor, but the urban population is likely to continue to increase as people without land or jobs are driven into cities.

Strain on resources

Bangladesh is the world's ninth most populous country, with about 123 million people according to official estimates. The Ministry of Health and Family Welfare says the population is expected to grow to about 210 million by the year 2020.

Despite the relative success of family planning programmes which have brought the population growth rate to between 1.6% and 1.8% a year, already overcrowded Bangladesh will inevitably suffer further strain on its limited resources and land.

The population is expected to stabilise at about a quarter of a billion by 2050, if family planning programmes continue to reduce the birth rate.

But Bangladesh is already one of the world's most densely-populated countries, and its low per capita income is unlikely to increase.

"In a small country like Bangladesh, it will be a really big problem to accommodate such a huge population," said Mizanur Rahman of the Family Planning Association of Bangladesh.

Millions migrate to cities in search of work. Malnutrition levels are already high. At least half of Bangladesh's people live below the poverty line. They eat only about half of the amount of food that would be considered normal elsewhere in the world

Bangladesh is already unable to provide jobs, housing and food for all its people, and some analysts fear that there may be an increase in crime and violence as increasing numbers of people have to compete for resources.

TASKS

1 Imagine you are an aid worker and that you have just made a field visit to the city of Dhaka in Bangladesh. Use the resources on this page to write a report outlining the problems created by the growth of squatter communities in cities in LEDCs.

2 Read the BBC news articles 'Deathtrap cities keep growing' and 'Bangladesh struggles to cope'. Explain the problems of urbanisation faced by the country.

The USGS website, Earthshots: Satellite Images of Environmental Change, includes an excellent case study about the impact of urbanisation on Santiago, Chile, in South America, including satellite images, photographs and text:

http://edcwww.cr.usgs.gov/earthshots/slow/tableofcontents

3 Produce a report, using word processing software, to explain the impact of urbanisation in Santiago and its surrounding area. Download data form the website and insert it in your report. You could, for example, include satellite images comparing the sizes of Santiago in 1975 and 1989, using the software tools to label changes.

SQUATTER SETTLEMENTS

Squatter settlements have been viewed by governments as places of squalor and despair. This is the view held by the Bangladeshi government in the news article from the *Daily Star Bangladesh*.

FIGURE 2.37

Squatter settlement problems

They see the answer to the problems of squatter communities in terms of bulldozing them and encouraging the residents to return to the countryside. Elsewhere governments and aid agencies are beginning to see these communities more positively. When viewed in the longer term, shanty towns pass through a series of stages of development. Initially they are indeed squalid and disorganised. With the passage of time, however, they can evolve into well-serviced, well-built suburbs.

The inhabitants of the squatter communities are often ambitious, hard-working people who showed the commitment and initiative to leave the rural areas to start a new life in the cities. Over the years these people establish themselves in the city and possess the drive and energy necessary to improve the squatter communities. Many of the inhabitants ultimately manage to get jobs with regular wages. Gradually they are able to buy better building materials to improve their homes. It may take up to 20 years but eventually a flimsy shack can be transformed into a well-built, fully serviced house.

Governments are coming to realise that, with support, squatters are capable of establishing communities. They cannot, however, afford to provide the infrastructure. The World Bank and leading charities are funding site and service schemes. The authorities provide a site, basic building materials such as breeze blocks, water, sewage systems and electricity. The local people build single-storey homes with a water tank. As the site develops, roads, clinics and schools may be added. The fact that the residents work together on these projects helps to develop a community spirit.

In Calcutta, India, the Metropolitan Development Authority was set up in 1970 to improve the quality of bustees, by paving alleys, digging drains and providing more water taps and toilets. In Cairo, Egypt, the authorities have attempted to improve the infrastructure, for example by organising refuse collection by licensing donkey carts. Much larger schemes involve the building of satellite towns away from Cairo.

TASKS

1 Using the information provided on these two pages describe ways in which the conditions in squatter communities can be improved.

2 Read the article (opposite left) from the *Daily Star Bangladesh*, 19 August 1999.

   Explain what the Bangladeshi government is trying to achieve by this scheme.

3 Read the article (opposite right) in which the World Bank criticises slum clearance.
   a Explain why the World Bank disagrees with the Bangladeshi clearance plan.
   b What solutions do they offer?

Measures to rehabilitate slum dwellers

Daily Star Bangladesh August 19, 1999

Prime Minister Sheikh Hasina yesterday announced a series of measures for rehabilitation of the slum dwellers in their respective village homes, report agencies.

Each family intending to return to their village will be provided with travel expenses, three months' free food support under VGF programme and micro credit facilities through various institutions and agencies of the government.

The Prime Minister announced the programmes while addressing first meeting of the recently constituted Slum Dwellers Rehabilitation Committee.

The Prime Minister referred to the government's massive programme for providing shelter and employment in the rural areas and said, 'We are working to ensure home for the homeless and better living for the people.'

The slum dwellers are leading a subhuman life in the city slums and are being exploited by a section of people, she said, adding that some people are using the city slums as haven for crimes.

The Prime Minister said the government is running the Asrayan Project for the homeless, a housing project for low-income group people, Adarsha Gram (ideal village) project and providing loan and training for the rehabilitation of the poor.

'We have provided homes to over 7500 families through the Asrayan project, about 12000 families through Adarsha Gram project and taken a new programme for providing homes to the low income group people through the housing project,' she said.

'The Krishi Bank is running a project to provide loans and shelter to the people who are returning back to the village', she said, adding that under this programme 'Ghare Fera' (back to home), already 2000 families have returned to the village and the programme is going on.

She said government has also taken steps to strengthen the rural economy and local government bodies through decentralisation of administration. She also mentioned other programmes for increasing facilities in the rural areas.

BBC News
Sunday, August 29, 1999
Published at 04:31 GMT 05:31 UK
World: South Asia

WORLD BANK CRITICISES SLUM CLEARANCE

By Kamal Ahmed in Dhaka

THE WORLD BANK'S SENIOR OFFICIAL IN BANGLADESH, Fredrick Temple, has criticised a recent slum eviction drive ordered by the government.

'Bulldozing slums was not a solution to the problems of urban life,' Mr Temple said.

'Forcible eviction without relocation simply shifted poor people from one set of slums to another,' he added. The World Bank director called for a national urban strategy to cope with the pressures of urbanisation, which he said was a result of economic growth centred around city life.

POLICY CRITICISED

Mr Temple's comments are the most explicit criticism so far from the donor community of the controversial slum demolition drive, which made at least 50,000 people homeless.

The World Bank is one of Bangladesh's largest foreign aid donors, providing about $1bn last year.

He said that, in general, the clearance programme – carried out because the government said some slum areas in Dhaka and its outskirts were centres of crime – did not constitute a full urban settlement or shelter policy.

'The experience of other countries indicates that it is neither possible, nor affordable, for cities with large areas of slums to relocate the inhabitants' he added.

MIGRATION

Quoting the latest Bank study, Mr Temple said that, while four out of five Bangladeshis live in the countryside at present, in two decades every other Bangladeshi would live in cities.

The World Bank director said he shared the concerns of the Bangladesh government on urban law and order. But the forcible eviction of slum dwellers undermined the Bank's efforts, in conjunction with the Bangladeshi government, to provide education, health-care, job-training and micro-credit to the poor.

Police launched the slum clearance operation earlier this month, saying it constituted a crackdown on crime.

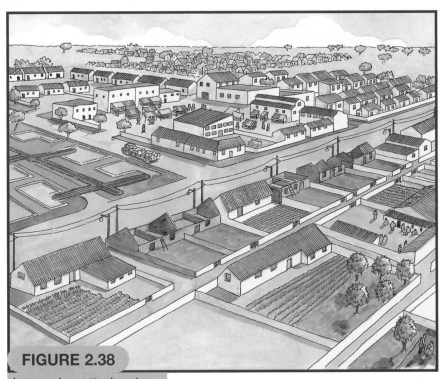

FIGURE 2.38

Improved squatter housing

THE CENTRAL BUSINESS DISTRICT

The Central Business District (**CBD**) is a small but very **accessible** area. Everyone living in the city has a fairly direct route into the centre. Most of the bus routes run directly to and from the CBD. There are so many shopkeepers and businesses who would like to use a building in the CBD that the cost of land (**the bid price**) here is very high. The best location for a shop will have the highest bid price. This is the Peak Land Value Point (PLVP). It is likely to be used by a chain store such as Marks and Spencer. Land in the CBD is too expensive to be used for houses and factories. Since land here is so scarce, the buildings in the CBD are often higher than anywhere else in the city. The floorspace is much greater than the groundspace, so the **effective area** of the CBD is much larger than it looks on the map. Land to use for car parks and bus stations is difficult to find.

Within the CBD, people usually walk from place to place. This helps to keep the CBD compact and encourages some functions to cluster together. There may be a financial part of the CBD, for example, containing most of the banks and insurance offices. Entertainment may cluster, with restaurants close to cinemas, theatres and night clubs. There will be important public buildings, such as the town hall and cathedral, but very little open space. There are some businesses which require a large area, yet want to be as close as possible to the centre. They cannot afford the bid price at the centre of the CBD, so they locate near the CBD's edge, where land is cheaper.

TASKS

1 What does **accessible** mean?

2 What is the advantage to a shopkeeper of having a shop at the most accessible point in a city?

3 Look at Figure 2.39. Swansea's CBD appears to cover half a square kilometre. On average, though, each building has three floors. What is its effective area?

4 Explain why there is very little open space within the CBD.

5 For journeys within the CBD, why do people walk rather than drive or use public transport?

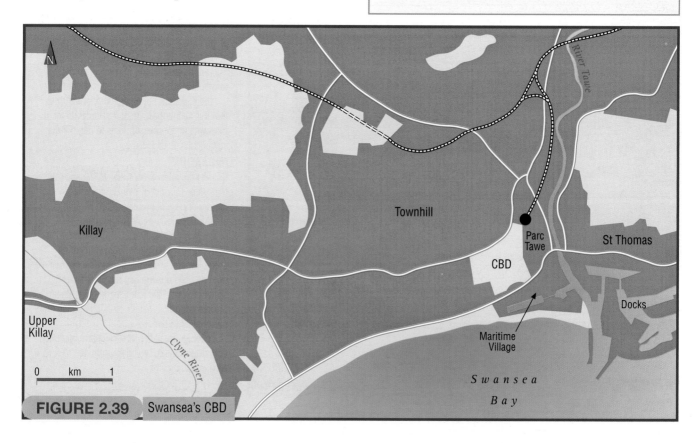

FIGURE 2.39 Swansea's CBD

Swansea's CBD

The CBD of Swansea contains about 450 shops. Shoppers come from a large area. It is the regional centre for the whole of south west Wales, and has department stores, the largest market in Wales and many small, specialist shops. Some **high order goods** are not available anywhere else within 50 km. The oldest commercial area is High Street, which runs south from the railway station. To the west are the largest shops, rebuilt after bombing during the Second World War. Most of this area is now **pedestrianised**. It includes The Quadrant Centre, opened in 1979, which is undercover and linked to a multi-storey car park.

FIGURE 2.40

Swansea's pedestrianised CBD

FIGURE 2.41

Out of town shopping at Llansamlet

Four kilometres north of the CBD is Llansamlet Retail Park, with 40 large stores. This part of the Lower Swansea Valley was once the world's greatest copper producing area. The last works closed in 1980. It became polluted waste land and an area of high unemployment. To encourage firms to create jobs here, the government made it an Enterprise Zone. Firms could move in quickly, as there was no need to wait for planning permission, and cheaply as they did not pay rates for ten years. There is plenty of parking and it is free.

One-third of all spending by shoppers in the UK is now at out-of-town centres. In many cities, out-of-town shopping has become the alternative to the CBD, where there are now fewer shops. Some of these have been replaced by charity and second-hand shops. In Swansea, though, the CBD is fighting back. More streets are now traffic free (Figure 2.40); Castle Gardens has been extended and planners are trying to introduce housing into the city centre, particularly on the fringe and above shops.

TASKS

6 What land uses can you identify in Figure 2.40? What suggests this is the CBD?

7 Describe how shopping centres have changed from streets busy with traffic to new developments today.

8 Some CBDs are empty soon after the shops close. What is Swansea doing to avoid this?

9 Would you like to live in a CBD? What are the good and bad points?

FIGURE 2.42

St Thomas, Swansea

THE INNER ZONE

Before the Industrial Revolution, Swansea was a
trading port at the mouth of the River Tawe. Its
population was 6 000. As the area developed into
the world's major producer of copper, new docks
were built and the population grew twenty fold.
Workers came for jobs in the new industries based
on metals and coal.

There was a great need for more housing. Areas
close to new industries and the docks, such as St
Thomas (Figure 2.42) were built. Many of these
buildings are now old, and many were not built to
last so long. Most of the houses are terraced, with
little or no front garden. They have two bedrooms
upstairs and two rooms below. Often the ground
floor extends back into the garden, to provide a
kitchen, coal shed and toilet.

Some of these houses were demolished where it
would cost too much to bring houses up to modern
standards. But many terraced houses in St Thomas
now have a toilet and bathroom inside, often as an
extension on top of the kitchen and coal shed.

FIGURE 2.43

Flats in Swansea

FIGURE 2.44

Maritime Village from the air

In some parts of Swansea's inner zone, terraced houses have been replaced by flats (Figure 2.43). When they were built, they seemed to offer much better living conditions but they are not popular with families. They lack gardens and garages; some are noisy, lifts are sometimes vandalised and residents can feel trapped. There are few places for children to play in safety and it is harder to build a sense of community than in an area of terraced housing.

Two areas of Swansea's inner zone have seen major changes. Between the CBD and the river is the Parc Tawe development (in the background in Figure 2.42). It is a mixture of shopping and leisure uses, such as a cinema, ten-pin bowling and Plantasia, a giant glasshouse containing a rainforest experience. Parc Tawe Phase 2 is redeveloping the inner zone to the east of the river.

Between the CBD and the beach was South Dock. A century ago it was a prosperous area of ships, warehouses and coal hoists. When the dock closed in 1969, Swansea Council bought it. It has now become the Maritime Village, with hotels, shops and restaurants (Figure 2.44). There are 1500 new homes, mostly apartments overlooking the old dock, which is now a yachting marina. One of the old warehouses has been turned into the Maritime Museum. Next to it, a new Leisure Centre has become the most visited tourist attraction in Wales. So far, about £500 million has been spent on the Maritime Village, including some grant aid from the European Union.

TASKS

1 Describe a typical terraced house in the inner zone.

2 What was the advantage of living here when the inner zone developed?

3 Look at Figure 2.42. What suggests this is part of the inner zone?

4 Describe the advantages and disadvantages of living in the flats in Figure 2.43. Consider:
a a family with young children
b a young single person
c an elderly couple.

5 Describe the redevelopment of the inner zone at Parc Tawe and the Maritime Village.

6 a If you live in a town or a city, with a digital camera take pictures of different land use zones; include the CBD, inner city and suburbs.
b Download the images and save them into a folder either on your computer or or a floppy disc.
c Insert your digital images into a desktop publishing program. Use the software's tools to label the characteristic features of each land use zone.

SUBURBS

Half the population of the United Kingdom lives in **suburbs**. They are the areas of houses and gardens of similar size and type adjacent to the city. The density of housing is medium to low. Many suburbs were built in the 1920s and 1930s on land which was then at the edge of the city. Estates of semi-detached houses were a great improvement on the back-to-back terraced slums of the inner zones. Many people moving to the suburbs had an inside toilet and bathroom for the first time. Although the docks and industries were now further away, it was worth the time and cost of the bus journey to work.

The Swansea suburb of Townhill was built by the council in the 1920s on steep slopes. Now long-term unemployment here is high at about 45%. Townhill has a young population. Almost half of the 13 500 people who live here are under 30. 40% of households with children, though, are single-parent families, most relying on state benefit.

There are few places for children to play and crime rates are high. School truancy is the highest in Wales and many 16-year-olds expect never to have a full-time job.

Now, however, £9 million is being invested in Townhill, half from the European Union. Local community groups have organised the 'Greening Programme' to improve the environment. It has provided work for a task force of five local people and increased pride in the community. Some money has been spent on the road network. Speed humps and chicanes slow the traffic; new pedestrian crossings and better street lighting make Townhill safer. To reduce unemployment, local colleges are offering training and providing crèches. There are grants for local businesses which promise to give job interviews to Townhill residents. There are plans to build a centre to give business advice and support local employment on Paradise Park on the top of Townhill. It is one of the few flat sites and is used for recreation.

FIGURE 2.45

Townhill shops

To control the growth of cities, the government set up **green belts**. These are areas in which new building is not normally allowed. Sometimes planners have permitted certain villages within the green belt to expand to provide more homes, but rebuilding the inner zone is the first choice.

In many cities, the suburbs are showing their age. The houses are decaying. There are few open spaces and local shopping centres are run down because shopping and leisure have moved out of town. With more people owning cars, there is less need to provide facilities like health centres in the suburbs and it is harder for local shopkeepers to make a profit. Suburban stores decline and close. The neighbourhood loses its sense of community. Much money has been spent on inner zones, but the suburbs have not been seen as priorities. Now there is **de-suburbanisation** as people move to places outside the city, leaving problem areas behind but creating pressure on rural areas.

FIGURE 2.46
Semi-detached Killay

The suburbs on the edge of the city include houses built more recently than Townhill. Killay (Figure 2.46) was originally a separate village. Its shopping centre offers a bank and travel agency, as well as shops for low order goods. Most houses in Killay are privately owned, have garages and open-plan front gardens.

Across the Clyne valley from Killay is Upper Killay, 1 km away. By road, it appears to be a continuous settlement because there is a line of houses on each side of the road. Behind them, however, is farmland. This is **ribbon development**. Planning laws have now stopped tentacles of housing developing into the countryside, along the main roads out of cities like Swansea.

TASKS

1 When the suburbs were built, what were the advantages and disadvantages of moving from the inner zone to the suburbs?

2 List some of Townhill's problems.

3 How is the quality of life being improved in Townhill?

4 Look at Figures 2.45 and 2.46. In what ways is the suburb of Killay different from Townhill?

5 Before planners stopped ribbon development, why was it attractive to builders to put houses alongside main roads?

6 Should new houses be allowed in green belts? Consider allowing cities to expand outwards from the suburbs, or green-belt villages to grow into small towns, or building on segments of green belt, leaving only green wedges.

TRAFFIC IN AN URBAN AREA

Most towns grew because they were **accessible** places. Glasgow in Scotland was the lowest bridging point on the River Clyde, where routes met to cross the river. It became a trading centre as well as a major port. Ships no longer come up river to the centre of Glasgow, but the roads are still busy – too busy. **Congestion** often delays traffic, fumes pollute the air and road accidents happen.

In the past, policies have aimed at improving traffic flows. But new roads and better junctions encouraged more traffic, which led to more congestion. Bus journeys became slower and more unreliable, air and noise pollution and accident levels increased. As a result, more people switched from bus to car, and parents now drive their children to school rather than letting them walk.

Glasgow has strategies to favour public transport, cyclists and pedestrians, while still allowing cars and commercial vehicles into the city. In ten years, however, the use of buses reduced by 13%. Now bus travel is being made more attractive. All parts of the city are within 300 m of a bus stop. Eighteen routes into the centre of Glasgow have a bus every ten minutes and every house has had a bus map delivered showing routes and connections. Some new 'bendibuses' can each carry 120 passengers and are fitted with traffic light transponders which turn the lights green as the bus approaches road junctions.

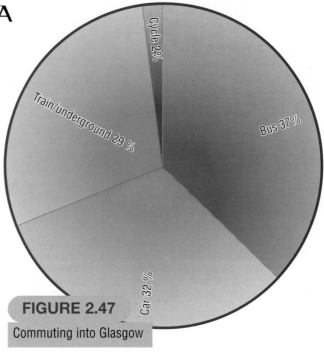

FIGURE 2.47

Commuting into Glasgow

Cycle 2%
Bus 37%
Train/underground 29 %
Car 32 %

Maryhill Road (Figure 2.49) was chosen for one of the first Route Action Plans. Where it is wide enough, a lane in each direction is a bus lane. Buses, taxis and cyclists can overtake queuing traffic. The plan is to reduce bus delays by 30%. New bus shelters have information signs showing how many minutes until the next bus arrives. In the city centre, some junctions allow buses through but not cars and delivery vehicles. These are called **bus gates** (Figure 2.48).

FIGURE 2.48

Bus gates speeding up public transport

FIGURE 2.49
Maryhill Road bus lane

Integrated public transport

Buses and trains link together in Glasgow. Apart from London, Glasgow has the most extensive local rail network in the UK. It is also expanding: new routes and stations have been opened. Now there are plans for a line to Glasgow Airport and a link across central Glasgow to join together train services north and south of the River Clyde. The plan is to have every part of the city within 500 m of a station and to make public transport more attractive by integrating train, bus and underground timetables, fares and tickets. One success has been **park and ride**. Most rail stations have free car parks (Figure 2.50) and there are now more than 5 000 parking spaces. The cost of providing them is soon met by ticket sales. It is even cheaper to provide **kiss and ride**, where a commuter is dropped off at the station and the car is driven away.

FIGURE 2.50
Park and ride rail station

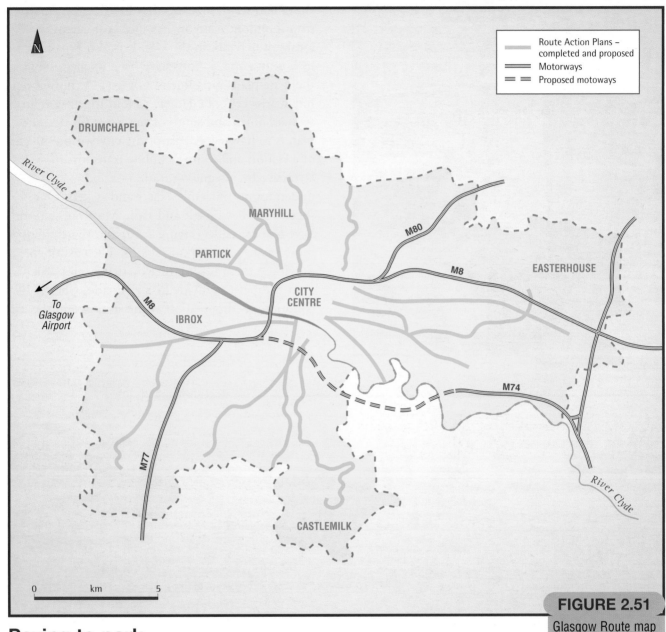

FIGURE 2.51

Glasgow Route map

Paying to park

Alongside making public transport more attractive is the plan to reduce commuting by car to less than a quarter of the cars parking in Glasgow. Within the city centre, the number of parking spaces has not increased in the last twenty years, but prices have. Although the cost of short-stay parking has gone up at double the rate of inflation, all-day parking costs have increased five fold! This has discouraged commuting by car to the city centre but favoured short-term shopping and business use. Many workers who used to drive in now use public transport but some park just outside the CBD and walk in. As a result, the council has extended the controlled parking area into the inner zone. People who live there now need permits to park outside their houses and the permits are not free.

Encouraging cycling

Glasgow has a cycle network of 100 km, mostly along closed railway routes. Cycling is expected to quadruple by 2012. Cycles can be taken on trains for free and bike lockers are being provided at rail stations, some bus stops and main shopping centres. As well as reducing car traffic, cycling keeps Glaswegians fit.

FIGURE 2.52

The busiest bridge in Britain, Kingston Bridge, carries the M8 motorway over the River Clyde in Glasgow

Moving goods

97% of Glasgow's freight travels throughout by road. Even the rest finishes its journey by road from the railfreight terminal or airport. To improve the quality of life, no deliveries are allowed in the city centre during the working day and there is a peak hour restriction on the main roads into the city to reduce congestion.

More roads

As part of the redevelopment of Glasgow's inner city in the 1960s, a motorway, the M8, was built round the north and west side of the CBD (Figure 2.52). It provides a fast route almost to the centre of the city. There is a proposal to extend Glasgow's motorway system along the south side of the River Clyde to bring the M74 to link to the M8 so that a motorway completely circles Glasgow.

TASKS

1 What two means of transport made Glasgow a good site for a settlement to develop?

2 There used to be a policy of providing for more traffic as traffic increased. Why has the policy changed?

3 Look at the pie chart of how people commute into Glasgow (Figure 2.47). Describe what it shows.

4 A common reason given for not choosing to travel by bus is its image.
   a List the 'good' and 'bad' points about bus travel.
   b What is already being done to improve the image?
   c Glasgow is planning to introduce electric buses. How will they change the image?

5 a What is the policy on car parking?
   b There are 7 800 spaces in public car parks, 3 500 on-street and 16 000 company car park spaces. How does this make the policy on commuting by car difficult to achieve?

6 Describe what has been done to encourage cycling in Glasgow.

7 Will more motorways help or make more difficult Glasgow's plans for public transport? Write down your reasons.

THE PROVISION OF SERVICES

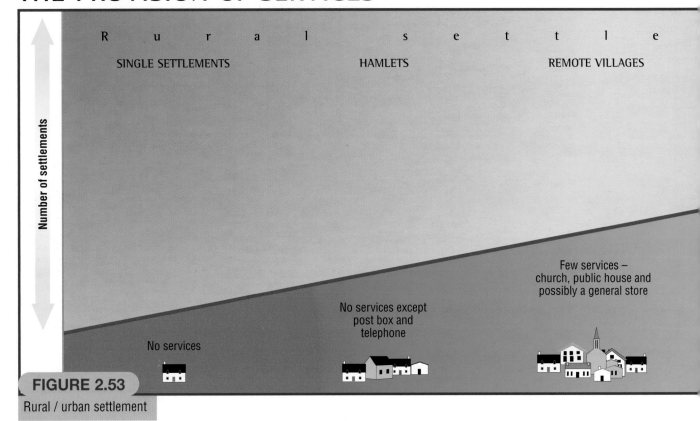

R u r a l   s e t t l e

SINGLE SETTLEMENTS

HAMLETS

REMOTE VILLAGES

Number of settlements

Few services –
church, public house and
possibly a general store

No services except
post box and
telephone

No services

FIGURE 2.53

Rural / urban settlement

The hierarchy of settlements

In every region, there are settlements of different sizes. The smaller the place, the more of them there will be. The smallest are **single settlements**, such as farms. Where a few homes are clustered together, the settlement may have a place name – and not much else. This is a **hamlet**.

A **village** is larger, with enough people to support a few services. Many villages which are accessible to urban areas have grown, as new houses have been built for people who moved out to live nearer the countryside (see page 105). Here a few specialist shops may be profitable, such as a hairdresser. Villages which are far from employment in urban areas have few commuters. Most have not grown. They have fewer services. Many do not have a shop.

For the goods and services not available in the village, residents must travel to a **town**. Here, a specialist such as a dentist has enough custom to be busy, serving both the residents of the town and those of the smaller settlements around.

The largest towns are often called **cities**. In every region, there is one major settlement providing the very high order goods and services, which require large **threshold populations**, such as an airport or television studio.

Towns and cities – too successful?

Places grew because they were accessible for trade. There was money to be made, but as more people wanted to set up business there, the price of the land rose. Cities became expensive locations, and less accessible for people and goods too, as congestion grew (page 105). **Out-of-town shopping** developed where land was cheaper and easier to get to, such as by a motorway junction. Free parking could often be provided. Not only has custom been taken away from town centres and neighbourhood centres in the suburbs, but the hypermarkets have meant fewer visits to the village shops.

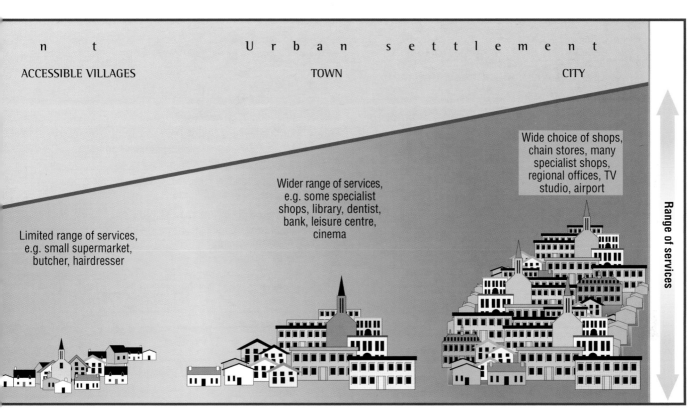

n t   U r b a n   s e t t l e m e n t

ACCESSIBLE VILLAGES   TOWN   CITY

Wide choice of shops, chain stores, many specialist shops, regional offices, TV studio, airport

Wider range of services, e.g. some specialist shops, library, dentist, bank, leisure centre, cinema

Limited range of services, e.g. small supermarket, butcher, hairdresser

Range of services

RURAL SERVICES FIGHT BACK

A quarter of the UK's population lives in rural areas – and the number is growing. Yet many rural services have been declining for years. Now there are signs of hope.

▶ Village shops
  – selling lottery tickets has brought in more customers,
  – many offer more personal services than hypermarkets such as home deliveries,
  – some councils help by charging lower rates, making loans available and giving free business advice,
  – village post offices are diversifying, such as selling travel insurance. In some villages, the post office is now inside the village hall or part of the pub.

▶ Public transport
  – community transport schemes such a 'dial a ride' are increasing, using volunteer drivers,
  – 'post buses', where a minibus rather than a van is used by Royal Mail, provide transport for local people,

  – some rural railway stations are reopening.

▶ Community facilities
  – many village halls have been improved,
  – primary school facilities are used by local people,
  – some pubs have been made into family pubs, with restaurant and play area.

▶ Mobile services
  – milk deliverers now also sell groceries,
  – as well as the library van visiting villages weekly, an increasing range of services, such as banking, video rental, crèche and butcher is now provided by van.

TASKS

1 a Which of the following could be provided in villages by mobile services?

  fish and chips, playgroup, dentist, church.

  b What problems need to be overcome?

2 You are the Chief Planner for a rural county. Produce a report to show how services in rural areas can be provided.

NORTH WORCESTERSHIRE, A CASE STUDY

Cookley is a village in north Worcestershire with a population of 2 500. Very few work there. Most commute to jobs in towns and it is there that most of their shopping is done.

Cookley only has a few shops and services (Figure 2.54). They provide mainly low order goods and services. Very few residents do all their food shopping in the village because prices are higher than at supermarkets, which also offer more choice. Cookley's shops are used for 'topping up'. The shop used by more Cookley residents than any other is the fish and chip shop. The village has three public houses, more than many villages of this size. Although pubs do have a low threshold population, Cookley's rely on custom from nearby towns to be profitable.

TASKS

1 Suggest why the fish and chip shop is used by more Cookley residents than any of the other shops in the village.

2 Work out the 6-figure grid reference for the centre of Cookley on the Ordnance Survey map opposite.

3 The size of print used on Ordnance Survey maps usually indicates the kind of settlement. Towns are in capital letters, villages in bolder lower case and so on.

  a How many towns are shown on the map opposite?
  b Name four villages shown on the map.
  c Look at the single settlements such as farms shown on the map. Are there more single settlements than villages? What is the relationship between the size of settlements and their frequency?

FIGURE 2.54

Cookley village

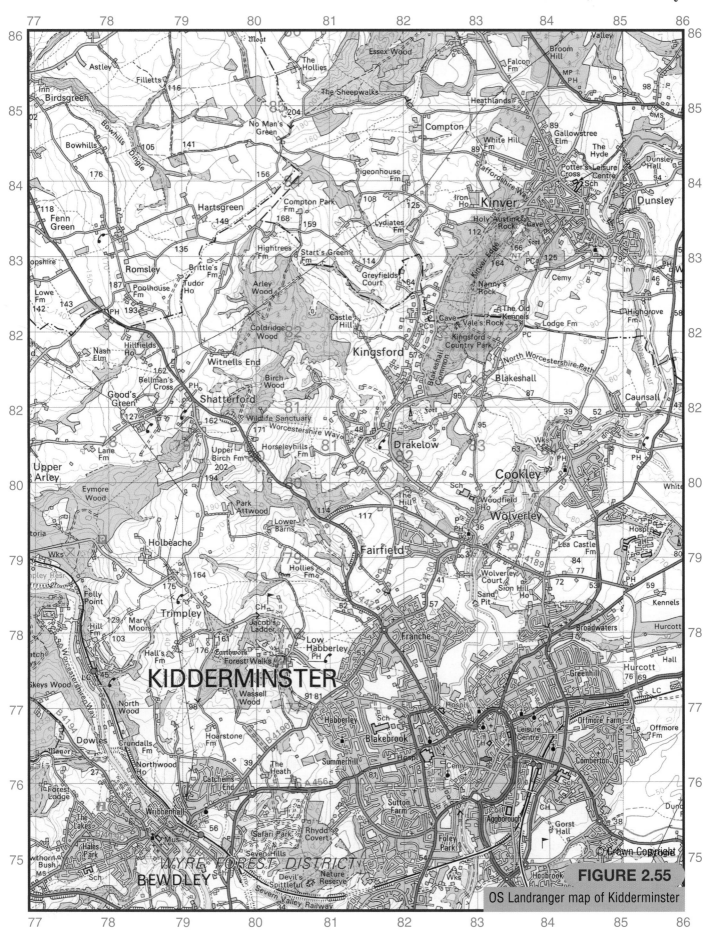

FIGURE 2.55

OS Landranger map of Kidderminster

Shops and services reflect the need for them, so, as car ownership has increased, the bus service has declined, with too few passengers to pay its way. Once there were two primary schools in the village but the families who moved into the new houses 40 years ago are now mature. With fewer children in Cookley, one of the schools has closed. The Parish Hall, however, is doing well with community activities for older people, such as barn dances, and the doctor's surgery is being extended. Older people make more use of medical services.

FIGURE 2.56

Offmore neighbourhood centre in Kidderminster

Much newer housing has been built in Cookley, particularly during the 1960s and 1970s. Although the population grew, shops and services did not increase. Almost all the new residents have their own transport and many choose to go to Kidderminster, the nearest town. Kidderminster provides goods and services to its own population and a large rural area which includes Cookley. An example is an estate agent. If the office were in Cookley, rather than Kidderminster, there would not be enough customers to be profitable.

Within towns, there are small clusters of shops away from the CBD, similar to those found in villages like Cookley. For low order goods, like newspapers, **neighbourhood centres** such as in the part of Kidderminster called Offmore (Figure 2.56), provide the everyday needs of people who live locally. High order goods have a larger threshold population, so a visit to the CBD is needed. For the highest order goods and services, even Kidderminster does not have a large enough sphere of influence. It is necessary to go to Worcester or Birmingham for these.

In the last 20 years, a new out of town centre has developed, 14 km from Kidderminster at Merry Hill. It had been the site of a steelworks. The wasteland was turned into undercover, air-conditioned malls with a mixture of department stores and smaller shops, a crèche, fast food, a multiscreen cinema and free parking for 10 000 cars.

To be successful, Merry Hill needed to attract shoppers from a very large area so that its high order shops have enough customers. $4\frac{1}{2}$ million people live within its sphere of influence. Some of the shopping which used to be done in Kidderminster (and even Cookley) is now done at Merry Hill. With fewer shoppers in Kidderminster, some high order shops, such as fashion shops, have closed down.

TASKS

4 Look at the OS map on page 103. Kidderminster is a route centre.
  a How many main roads meet there?
  b What other types of transport are shown on the map?

5 a Suggest the grid square on the OS map which includes Kidderminster's CBD.
  b Why did you choose that grid square?

6 Look at Figure 2.56. What can be bought at Offmore neighbourhood centre?

7 What are the attractions of shopping in a large out-of-town shopping centre, such as Merry Hill?

8 What is the difference between high order and low order goods?

9 What problems do shopping centres in the middle of towns face? Think of a town you know well. What has been done to make it an attractive place to shop?

10 Think of a large, modern out-of-town shopping centre that you know. What are the advantages and disadvantages of shopping there?

URBAN TO RURAL MIGRATION

FIGURE 2.57

Sheep for neighbours in Cutnall Green

Towns grow where there is money to be made. In the UK, rapid town growth began with the Industrial Revolution. Towns on, or close to coalfields were profitable places for metal-making and metal-using industries. People left the countryside to earn money in the towns. There was rapid **urbanisation**. This process continues in many LEDCs (pages 82–83).

In the UK, however, the population of almost all large towns and cities has been falling since the 1960s. More people are leaving the cities than moving in to them. This is **counterurbanisation**.

Causes

Just like migration to cities in LEDCs, counterurbanisation results from 'push' and 'pull' factors. The push comes from a shortage in cities of attractive homes at affordable prices, the wish to escape from traffic congestion, air pollution and sometimes, a fear of crime. The pull of the countryside includes the belief that the air will be clean and life will be less stressful, people will be friendlier and there will be a better environment for children to grow up.

Counterurbanisation has been made possible by changes in work patterns. Technology now allows more people to work from home by computer link. Some footloose industry (page 152) is choosing rural locations, which encourages urban to rural migration. For many people though, work is still in the city and a rural home means **commuting**. Improvements in transport have encouraged urban to rural migration. The cost of commuting can be thousands of pounds each year as well as several hours of travelling each day. Some people move from the city to the countryside to retire, often to live close to relatives or friends, or to places previously enjoyed on holiday.

The consequences of urban to rural migration

The effect on the countryside is greatest near to cities. Some villages have become dormitory settlements with the daytime population much less than at night. New housing developments have changed the character as well as the size of villages.

FIGURE 2.58

A second home by the River Severn

With a population similar to a small town, there are now supermarkets and some specialist shops. There may also be some offices and light industry, attracted by the pleasant environment and cheap land. The village is no longer a tight-knit community, and there can be tension between newcomers and the older established residents.

The countryside around the UK's major cities is protected from development as **green belt** (page 95). Villages within a green belt with their traditional church and pub may look unspoilt, but their character has changed.

Labourers' cottages and farm barns have been converted into luxury homes, too expensive for most local young people. They are likely to have to leave the village to find somewhere they can afford, usually in a town.

So urban to rural migration is partly balanced by the movement of people from rural to urban areas for work or education.

North Worcestershire

The villages of North Worcestershire are less than an hour's journey from Birmingham. Commuting to Kidderminster, Worcester or the southern part of the West Midlands conurbation takes even less time. This is an important 'pull' factor explaining why these villages are growing. Another is the price of houses. A house with four bedrooms in one of the villages is about 20% cheaper than the same size house in Solihull or Sutton Coldfield, similarly attractive residential environments just within the West Midlands conurbation.

TASKS

1 Page 83 shows reasons why people are moving to cities in LEDCs. Now draw a Push–Pull diagram for counterurbanisation in the UK. Annotate it to show why people leave the cities to live in the countryside.

2 a Why are areas close to conurbations often called 'the commuter belt'?

   b How has the growth of some villages in the commuter belt changed the life of the village?

   c Why have house prices risen in these villages?

3 Draw a diagram to show migration between villages and urban areas. Label arrows on the diagram to show the type of people who have moved to and from the villages. Write annotations to show the effects of migration on each place.

In the villages, the pace of life is slower and some people have achieved their lifelong dream by moving there. For some it is a second home on the bank of the River Severn. Many were built near Bewdley before planning laws were passed. Empty for much of the year, their owners make little contribution to the village communities. For others, after a lifetime of living and working in the conurbation, they have retired to North Worcestershire. In many villages, this has gradually changed the population structure. They now have a larger proportion of old people than towns like Kidderminster or Worcester and it is to the towns that many young people from the villages have moved to afford their first home. A two-bedroom starter home in Kidderminster is at least 20% cheaper than a two-bedroom cottage in a village.

Where the change in population has been gradual, new villagers have been more easily accepted. In some villages, though, new housing estates have been allowed. With rapid growth, 'incomers' have been less welcome and there has been less mixing.

Cutnall Green, 10 km south east of Kidderminster, had a population of 322 just before Brookfields Estate was added in 1997. Now an extra hundred people live in the village. The shop and post office, however, has not seen its trade increase by a third. The town of Droitwich is just 4 km away and provides a good range of shops. Most of the new arrivals come from towns and all but three commute to towns. Now 30% of Cutnall Green's population is aged 45 to 59, whereas for the whole of England it is about 17%. This reflects the **gentrification** of the village.

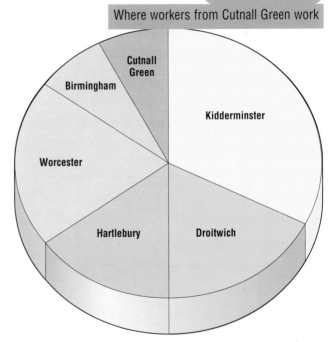

FIGURE 2.60

Where workers from Cutnall Green work

FIGURE 2.59

The Chequers Pub, Cutnall Green

TASKS

4 Suggest why house prices in North Worcestershire are cheaper than Solihull or Sutton Coldfield.

5 Describe the second home in Figure 2.58.

6 What are the attractions of living in one of the houses in Figure 2.57?

7 Imagine you have lived all your life in Cutnall Green. Describe the changes that have taken place in the village. Give your views on these changes.

COURSEWORK FOR SETTLEMENT

The settlement section of the syllabus is the one upon which the greatest number of coursework submissions are based. It makes good sense to investigate geographical patterns and issues which are local. Local knowledge is an advantage for coursework and it is easy to revisit study sites if additional data collection is needed. Of course, local opportunities for fieldwork vary greatly. For those living in a city there is too big an area to study and too much, while for village dwellers there may be too little so that their home village needs to be compared with a neighbouring village to allow sufficient data to be collected.

CBD studies

The CBD is the most distinctive of urban zones. It is a dynamic zone of constant change. These characteristics make it a suitable place for study. In small towns, such as rural market towns, it may be possible to undertake a whole CBD study, but in larger towns and cities it is essential to concentrate upon just one part or one theme.

Suggested titles

Note – where town has been used, this suggests that the title may better suited to a smaller urban area.

▶ What is the sphere of influence of the CBD of town A?

▶ What are the characteristics of the CBD of town B?

▶ What characteristics typical of a CBD are found in town C?

▶ Delimiting the CBD of town D.

▶ Where is the core of the CBD in city E?

▶ What changes occur as you move away from the core to the edge of the CBD in city F?

▶ Test the hypothesis that 'The core of city G's CBD is around the market place'.

▶ In what ways and why have land uses changed in the CBD of city H in the last ten years?

▶ What effect has the opening of the new indoor shopping centre had on the rest of the CBD in city I?

▶ An investigation into the traffic problems and attempted solutions in town J.

▶ An investigation into parking in the CBD of city K.

FIGURE 2.61

London's Oxford Street – which characteristics typical of a CBD can be seen?

Characteristic of the CBD	Method of data collection
1. Land use Land uses are dominated by commercial activities such as shops, offices, eating places and entertainments – but not houses.	Land use survey noting observations on a large-scale map or plan using a simple land use classification (Figure 2.63).
2. Shops Here are the largest shops, the widest variety of types of shop, and the shops with the largest threshold needing large numbers of customers to be profitable.	Land use survey as above but concentrating only on the shops (mainly A–F in Figure 2.63). Also – measure the size of shop frontages by pacing, record the number of storeys in use, count the number of customers going in and out of different shops.
3. Sphere of influence People will travel long distances to use the shops and services of the CBD. The larger the settlement, the greater the distance people will travel.	Design a questionnaire for shoppers to find out where they live, frequency of visit, mode of transport, shops visited, type of goods bought, etc. You can work out distance travelled and draw a desire line map for the sphere of influence (Figure 2.64).
4. Core and periphery A central area (core) is busier and looks more prosperous than the outer area (periphery), in which shopping quality and street appearance decline. There is a decrease in numbers of pedestrians.	Land use surveys (as above) with smaller shops, offices and public buildings expected in the periphery. Do an environmental quality survey using factors such as litter, care of the pavements, street furniture, traffic and noise. Conduct pedestrian counts.
5. Change This is a zone of constant change – in land uses, redevelopment of shops and offices, traffic schemes and pedestrianisation.	Compare current land uses with those shown on Goad maps from previous years. Compare old plans showing the layout of buildings and streets with present-day observations.

Figure 2.62 Investigating the CBD

Letter	Description of land use
A	Department stores and chain stores selling a variety of goods
B	Clothing and shoe shops
C	Specialist shops, e.g. books, sports goods, jewellers, florists
D	Convenience shops, e.g. food (bakers, butchers, sweets, etc.), newsagents
E	Furniture and carpets
F	Personal services, e.g. hairdresser, travel agent, dry cleaner, TV rental, etc
G	Catering and entertainment, e.g. eating places, pubs, hotels, cinemas
H	Offices and professional services, e.g. bank, building society, estate agent, etc
J	Public offices and buildings, e.g. town hall, library, main post office, job centre, etc
K	Transport, e.g. bus and rail stations, car parks
L	Change, e.g. vacant premises, building sites
M	Others, e.g. residential, industrial

Figure 2.63 A simple land use classification for CBD studies

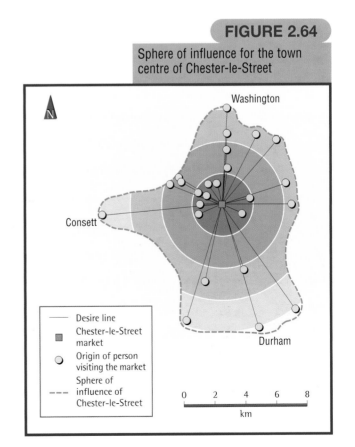

FIGURE 2.64

Sphere of influence for the town centre of Chester-le-Street

Washington

Consett

Durham

— Desire line
■ Chester-le-Street market
○ Origin of person visiting the market
- - - Sphere of influence of Chester-le-Street

0 2 4 6 8
km

Other urban zones

Only for a very small settlement can the whole of its built-up area be studied. The normal approach to land use investigations in towns and cities is to use one or more transects. Each transect begins in the centre and follows the course of a main road to the edge of the built-up area.

Although in small towns it may be feasible to use two transects to increase the coverage, in most towns sufficient data will be collected by making observations along only one transect. Indeed, in large towns and cities the transect may be so long that observations will need to be limited to every 20 m or 20 paces, or you may observe land uses on just one side of the road. Some suggested titles are given below.

▶ How and why do land uses change along road X between the centre and edge of town?

▶ Testing the hypothesis that 'The quality of the environment increases with distance from the centre'.

▶ Does the type, size, quality and cost of housing increase towards the edge of the built-up area?

Also popular are land use investigations that examine the links between urban zones and urban models.

▶ Does town Y fit the Burgess model?

▶ How closely do the urban zones of town Z match those of the urban models such as Burgess and Hoyt?

Steps in planning and completing a land use study related to urban models

1 Collect secondary information about the models from text books (but not too much).

2 Think about the urban land uses you are going to observe. Put them in classes. An example of a classification scheme is given in Figure 2.64.

AN URBAN LAND USE CLASSIFICATION SCHEME

Code number	Code letter
1 Housing	T = Terraced S = Semi-detached D = Detached F = Flats B = Bungalow
2 Shops and services	S = Shop B = Bank or Building Society O = Office G = Garage
3 Industrial	F = Factory W = Warehouse B = Builder's yard
4 Public buildings	S = School C = Church L = Library E = Other Educational such as a College
5 Entertainment	H = Hotel P = Pub E = Eating place L = Leisure Centre A = Arcade C = Cinema
6 Open space	CP = Car Park P = Park C = Cemetery S = Sports ground F = Farmland G = Gardens and allotments
7 Unused land	D = Derelict or empty building W = Waste land V = Vacant land being developed

Figure 2.64 An urban land use classification scheme

The numbers give you the function of the buildings such as houses or shops, and the letter gives you additional information, particularly for types of housing, which will help in deciding where to draw the lines between urban zones.

FIGURE 2.65

Suburban shopping centre. Is the land use classification in Figure 2.64 adequate, or do other land uses need to be added?

Advice

Do not automatically use the classification in Figure 2.64 or another book. Make changes to suit the needs of your own transect. For example, you may find it useful to distinguish several different types of shops, especially if your transect passes through several suburban shopping centres, such as the one shown in Figure 2.65.

3 From a map of the town/city, decide upon the route from the centre to the edge of town to be used for the transect.

4 Make a copy of the property blocks and streets along the transect from a large-scale map or plan.

5 Add the number and letter from the classification scheme onto your base map.

6 When the transect has been completed, decide upon the dividing lines between CBD, inner city, inner and outer suburbs and the green belt. In doing this bear in mind the characteristic features of each of the urban zones. Explain your decisions. Finally compare your zones with those of the urban models and suggest reasons for similarities and differences between your pattern of urban zones and those of the model(s).

Housing surveys

Some students are over-ambitious when they choose a large town or city for a transect which they try to relate to the pattern of zones in the urban models. Many would have been better off concentrating upon one type of land use such as houses or shops. Comparing two or more housing areas within the town or city can lead to good quality coursework provided that their locations within the built-up area are shown and commented upon in the Introduction and in the Analysis.

Useful points about doing housing surveys:
1 Varied and plentiful data collection:
▶ primary data collection by observation – house type, improvements, environmental quality survey;
▶ secondary data collection using house prices;
▶ primary data collection by questionnaires – socio-economic characteristics of the residents, such as age, type and place of work, habits, e.g. shopping, holidays;
▶ secondary data collection using the census.

2 Much of the fieldwork can be done locally. Common weaknesses of housing surveys done by students:
▶ few references to location within the urban area;
▶ no mention of urban zones and the geographical background;
▶ house prices stated but not used e.g. for averages in different housing areas.

FIGURE 2.66

What suggests this is part of the inner zone?

	Quality	Good	5	4	3	2	1	Bad
1	Housing layout and design	Varied and interesting Well spaced out						Poor and unimaginative High density
2	Building materials	Attractive						Drab and uninteresting
3	House maintenance	Well maintained, e.g. fresh paint, new doors and windows						Poor maintenance, e.g. peeling paint, broken pipes and gutters
4	Other features	Trees, grass and flowers improve the appearance						No trees, grass and flowers
5	Open space	Is present and there are safe play areas for children						Absent with nowhere for children to play
6	Gardens	Present and well looked after						No gardens or poorly maintained
7	Car parking	Parking mainly off the roads						Parking mainly on the roads
8	Road crossing	Little traffic, slow moving, easy and safe to cross						Busy roads, difficult and dangerous to cross
9	Litter	No litter visible						Heavily littered
10	Vandalism	No obvious signs						Graffiti, signs of damage and abuse

Figure 2.67 Measuring environmental quality of housing areas

Many students make good use of environmental quality surveys. Figure 2.67 is just one classification which should be adapted to suit individual need. If you feel strongly that something is good, tick 5. If it is quite good, tick 4 and so on.

Advice

The main weakness of environmental quality surveys is that they have to be based on subjective judgements. To overcome this, give some examples of your assessments based on photographs or sketches when writing up the method section. Look at the two housing areas in Figures 2.66 and 2.68. How many marks for environmental quality would you give each one? Check your total against those of your friends.

FIGURE 2.68

Housing area – what is it worth for environmental quality out of 50?

Surveys of shops and services

As with surveys of housing areas, it is usually best to compare two shopping centres so that you can describe, explain and comment upon the similarities and differences between them. The study of just one suburban shopping centre would need to be very detailed to yield sufficient data to allow work of grades A or B standard to be produced. Shopping centres form a hierarchy – they can be arranged in order of size and importance. Figure 2.70 shows a shopping hierarchy. The higher a shopping centre's position in the hierarchy,

▸ the greater the total number of shops;

▸ the greater the range and variety of shops and services;

▸ the greater its sphere of influence.

If you decide to compare two or more shopping centres, choose them from different positions in the hierarchy so that there is likely to be more upon which you can comment. However, it won't normally be possible to compare two if you are studying a CBD or out-of-town shopping centre at the top of the hierarchy. One will be large enough to give you all the data you need. The only way to keep comparisons to a reasonable size would be to compare certain types of shops and services at different locations such as in the CBD and in an out-of-town or suburban shopping centre.

FIGURE 2.69

Many new out of town shopping centres, which are privately owned, do not allow fieldwork without their permission

If you much prefer to collect data by observation than by asking people questions, concentrate on the number, range and variety of different types of shops. Also you can extend the shopping study to include traffic, parking and an environmental quality survey. If you are happy to stop shoppers and ask them questions, the sphere of influence theme may be the better option. There is a half-way house for those of you who don't like using questionnaires with the general public – devise a questionnaire about use of the shopping centres under study for people in your street and in the area where you live, who are not complete strangers.

CBD

Shops along main roads

Suburban shopping areas

Small corner shops

FIGURE 2.70 Shopping hierarchy

Village studies

If you live in a rural area, your opportunities for undertaking fieldwork are slightly different. The suburban areas of towns are gradually extending into countryside and you may wish to study a new development in the rural–urban fringe such as an out-of-town shopping centre, retail park, business park or bypass. First, however, you are likely to investigate the possibilities of doing fieldwork in the village where you live.

Some suggested titles are given below.

▶ What are the differences between the old and new parts of the village?

▶ Test the hypothesis that 'The new residents travel further to work and use the village services less than the old residents of the village'.

▶ An investigation into the changes which have occurred in the village since the mine closed in 1993.

▶ Does village A have the characteristics typical of a commuter village?

FIGURE 2.71

Entrance from the bypass into the village. What fieldwork possibilities lie ahead?

Possibilities

A Have there been recent changes in the village, whether growth or decline?

For example, many villages have grown after a new housing estate has been built so that differences between the new housing and the new arrivals on one hand and the old housing and long established residents can be investigated. In mining villages, there have been both human and environmental changes after the pits have closed.

B Is there a local issue?

Examples of these are the need for a bypass, or whether more farmland should be taken for new housing, or the problems caused by decline in local services.

▶ To what extent is village B now a commuter village?

▶ Does village C need a bypass?

▶ For a rural area, how is the provision of village services related to village populations?

▶ What are the arguments for and against the building of a new housing estate in the village?

If there isn't an obvious theme for your village, try a comparison of two villages with respect to layout and growth, which should make for a well-focused study.

Data sources for village studies

Primary data

Observation   – land use survey (houses, services, etc.)
             – environmental quality survey

Questionnaire – to all the different types of residents

Count   – traffic counts on different days at different times

Secondary data

▶ Old maps and photographs
▶ Parish records
▶ Census data
▶ Newspapers

The student's plan for data collection in Figure 2.72 shows that it is possible to collect a considerable amount of information even for a village. Why is the data collection suitable for the student's title? What shows that it has been well thought out? Data collection must always be carefully planned, because without sufficient data presentation and analysis also suffer.

Dangers associated with village studies:

A Not getting the balance right between primary and secondary data:

- going back too far with little about the village today;

- use of masses of secondary historical information.

B Not based on a theme or issue – just a village study.

Title 'Does my home village need a bypass?'

DATA COLLECTION PLAN

- Measurement along the main road through the village.
- Do traffic counts at different times of the day for one week.
- Measure noise pollution along the road using a sound meter.

OBSERVATION

- Do an environmental quality survey along the main road.
- Compare the results with surveys along other streets off the road.

QUESTIONNAIRE

- To 20 people living along the sides of the main road.
- To 20 people in other parts of the village selected randomly.
- A separate questionnaire to the shopkeepers and garage owner.

OTHERS

- Looking for articles in the local newspaper from the library.
- Visit the Parish Councillor.

Figure 2.72 Student's plan for a village study

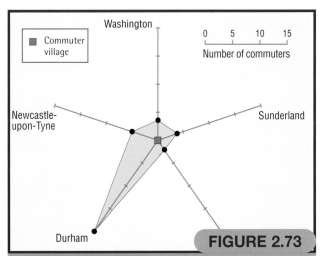

FIGURE 2.73

A star diagram showing where commuters from the village work – a presentation technique likely to be useful in many village studies

EXAMINATION TECHNIQUE – UNIT 2

At the end of your course, you will take either Papers 1 and 3 (Foundation Tier) or Papers 2 and 4 (Higher Tier). All four papers could include questions about **Population** and **Settlement**. Do not assume if you answer a Settlement question in Paper 1 or 2 that there will not be another in Paper 3 or 4. It is a major section, covering a lot of geography – enough content to test different topics in the different papers.

An important concept in geography is **distribution**. This word is included in the glossary on page 239. The questions to be examined first in this section focus on maps which show distribution. The following question is from the Higher Tier (Paper 2).

1 (a) Study Fig. 1a which shows the distribution of population densities around the world.

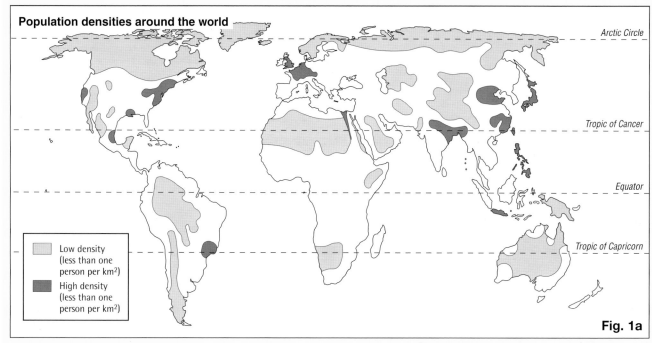

Population densities around the world

Arctic Circle

Tropic of Cancer

Equator

Tropic of Capricorn

Low density (less than one person per km²)

High density (less than one person per km²)

Fig. 1a

Briefly describe the world distribution of the areas of high population density (over 100 people per sq. km.)
[2]

The first points to notice are that one of the **command words** is 'briefly' and that the mark allocation is only two marks. Therefore a short answer is required but it must be precise. To focus your answer on 'distribution', think about the following questions:

▶ Where are the areas of high population density?

▶ Are these areas concentrated together or spread out?

Distribution does not mean give a list of the different locations which have a high population density. The other clues provided on the map are the three lines of latitude which can be used to describe distribution.

A concise answer which scores two marks is as follows:

Most of the areas of high population density are north of the Equator (or north of the Tropic of Cancer). These areas are spread out unevenly (or scattered around the world).

Do not write 'above the equator', it is a poor geographical description. Always use compass points when describing distribution.

The concept of population distribution is developed in the next question.

(b) Study Figs. 1a and 1b. Fig. 1b shows the **location** of four different types of environment. (If you are unsure what 'location' means, look it up in the glossary.)

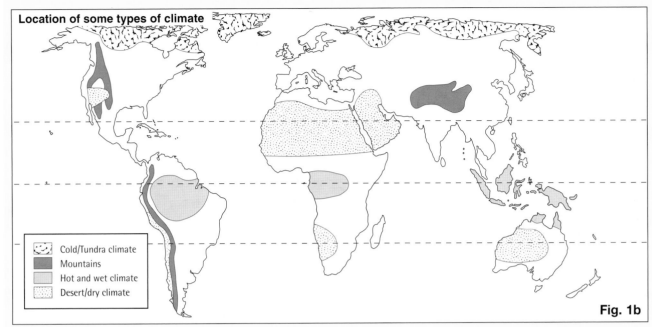

Location of some types of climate

Cold/Tundra climate
Mountains
Hot and wet climate
Desert/dry climate

Fig. 1b

Using Fig. 1b suggest reasons for the distribution of areas of **low** population density (less than 1 person per sq. km.). Develop your ideas fully. [7]

The map shows four different types of 'hostile environment'. The Higher Tier question tests your knowledge and understanding of why large numbers of people do not live in such areas.

The instruction to 'develop your ideas fully', and the large mark allocation, indicate that the answer should be wide-ranging and detailed, but not repetitive. The four different environments shown will all make farming, industrial location and the development of transport routes difficult. No credit will be given for including this sentence four times in your answer. It is not enough to state that an area is 'too cold or too dry for farming'. You need to explain why these conditions prevent large-scale farming.

The following is an example of a high-scoring answer. Note how the ideas are developed and the information from the map is used, not just copied.

Candidate's answer

The reason for low population density in North Africa is because it is a desert region with a dry climate. This shortage of water will make it difficult to grow crops which need regular

watering. The communication infrastructure will not be very developed over this large area and so it cannot attract people to live here.

So few people are found around the Arctic Circle because it has a tundra climate. There are no soils here to grow crops and the low temperatures and short growing season will make crop growing difficult. There is no communication here so industry, housing, and therefore people will not want to live here. The equatorial climate of the Amazon basin makes it very wet, hot and humid in this area. The area is tightly packed with trees so it would cost a lot of money to clear areas and build towns. The area is not desirable for most people to live because of the humid conditions and risks from tropical diseases and pests. The mountainous area of the Himalayas means that it is hard to construct roads, buildings and industry because of the steep slopes and high altitude. Most people will not be attracted to this area because of its lack of services and potential.

The **theme** of this question now has a change of emphasis from low population density to high population density. The following question is taken from the Foundation Tier (Paper 1).

(c) Name an area which you have studied with a high population density (over 150 per sq. km.). Explain why it is densely populated.[4]

The question is an opportunity to use a **case study**. One mark is reserved for a correctly named area. In your answer you must give some specific information about your chosen location. The answers below show how three candidates have tackled this question. Try to work out what makes the answers different in quality before you read the examiner's comments.

Answer from candidate 1

Named area: South Wales near Cardiff

It is very highly populated here because there was a lot of mining for raw materials like coal and iron, and many people moved here to work in the mines and ports where they now import raw materials.

Examiner's comment

The candidate names a good example, giving not only the name of the region but also a city within the area. Three reasons are given for the high population density, each one is accurate and specific to the region.

Answer from candidate 2

Named area: London
London is very crowded because of immigration. Many people move to the city to look for jobs and housing. But there are too many people. It is a large city so people believe that by moving, their standard of living will increase.

Examiner's comment

A correct example is named and the candidate then gives two reasons for the high population density. However, having referred to migration and jobs the answer becomes repetitive.

Answer from candidate 3

Named area: Birmingham
Birmingham is densely populated because people come to find work. It has a lot of shops and people want to be near shopping centres to buy their groceries.

Examiner's comment

The city is a correct example but only one vague reason is given for the high population density. All cities have shops which are built because lots of people already live in the area

The following question is from the Higher Tier (Paper 2).

(d) Explain why the rate of population **growth** is high in some LEDCs. [6]

The word 'growth' is in bold print to make sure that the candidate realises that a change of emphasis has taken place. Population growth is another important geographical **concept**. Answers should focus on birth and death rates and particularly the link between the two. The three answers again show different levels of success.

Answer from candidate 1

In some LEDCs the population growth is high because there is a high birth rate and a falling death rate. This occurs in LEDCs which are developing. They have high birth rates because they have little access to contraception, parents need children to help bring money to the family, parents need children to look after them in their old age, also the status of women is low as their role is to be mothers and not to have jobs. The death rate is getting lower because there is better sanitation, more hospitals, inoculations against disease and housing is less cramped. Also there are better food and water supplies.

Examiner's comment

The candidate shows an understanding that population growth depends on the birth rate being higher than the death rate. The candidate then goes on to give four reasons why there is still a high birth rate in many LEDCs, and finally explains why the death rate will decrease through improvements in many of the things which affect the quality of life.

Answer from candidate 2

The birth rate, which causes growth, is so high in many LEDCs because there is limited access to and knowledge of birth control. Some parents may not know about contraceptives because they do not have family planning clinics. Many people in LEDCs are subsistence farmers. This means that they will need a family big enough to help them in their work. The infant mortality rate tends to be higher in LEDCs. Therefore parents try for more babies on the presumption that there will be a higher chance of more surviving. As there are no pensions in LEDCs the parents will eventually need a large enough family to support them in old age.

Examiner's comment

This is an excellent answer to explain why birth rates are high in many LEDCs. Unfortunately the candidate only scores up to the maximum of four marks because there is no reference to the link between birth and death rates to explain population growth. There is also no explanation as to why the death rate may fall in such countries.

Answer from candidate 3

The rate of population growth is high in some LEDCs because families need extra workers to work on their farm, therefore they have more children. That may also get child benefits. The extra money that comes in from these children going out to work helps the parents a great deal.

Examiner's comment

This candidate only gives two reasons for high birth rates. Both ideas are linked to employment. This is a poor answer which does not properly explain population growth.

Questions from the settlement section of Unit 2 may focus on Papers 3 and 4. These are often referred to as the **'skills'** papers because most questions on these papers test your ability to interpret maps, photographs and graphs. However, do not forget that some questions on these papers will test your understanding of the **concepts** behind the geographical skills.

The following questions use the OS map extract on page 121.

> **2 (a)** Give a four figure grid reference for the grid square of the OS map extract which includes the CBD (Central Business District) of St Helens. Give one piece of evidence to support your answer.   [2]

At first glance this question appears to be testing your ability to give a four figure grid reference, but in order to interpret the map you must understand the concept of the CBD and the type of land use found in this area of a town or city.

> **(b)** What is the distance by road in kilometres from the junction of the M62 (505899) to the roundabout at 514948?   [1]

This question must be answered accurately by measuring along the road rather than the straight line distance. You do not have to use a ruler and then convert centimetres to kilometres. An easy technique is to mark the edge of a piece of paper along the route of the road, turning it as the route turns and then measure the length of the straight edge on the scale. In class, though not in the examination, a piece of string can be used.

The next question is on both Papers 3 and 4.

> **(c)** Use evidence from the OS map extract to suggest why Billinge (in and around Grid Square 5200) is a larger settlement than Houghwood (Grid Square 5100). [2]

The mark allocation suggests that two ideas are needed. Your answer must use map evidence (i.e. what you can see on the map). You will not gain any marks for an answer like the one on the next page because there is no evidence of this on the map.

Candidate's answer

Billinge is bigger than Houghwood because there are more shops and there is an industrial estate where the people will work.

Also remember that to answer this question you must **compare** two settlements. Therefore your answer must show this comparison, as illustrated below in an answer from a Foundation Tier candidate.

Candidate's answer

There are many main roads running through or adjacent to Billinge, whereas there is only one small road going to Houghwood so the transport is not as good. Also the land is flatter around Billinge than Houghwood, so farming would be easier there.

The following questions refer to five settlements located on the OS map. Information about their services is given in the table below.

Settlement	Grid Square	Population	Post Box	Church or Chapel	Public House	Post Office	Primary School	Hairdresser	Surgery/Health Ctr.	Dentist	Secondary School
Billinge	5200	9300	✓	✓	✓	✓	✓	✓	✓	✓	✗
Crank	5099	500	✓	✓	✓	✓	✗	✗	✗	✗	✗
Houghwood	5100	50	✗	✗	✗	✗	✗	✗	✗	✗	✗
Kings Moss	5001	150	✓	✗	✓	✗	✗	✗	✗	✗	✗
Rainford	4801	7100	✓	✓	✓	✓	✓	✓	✓	✓	✓

Key
✓ = Service present
✗ = Service absent

Notice how the questions based on this table are different in the Foundation and Higher papers.

Foundation Tier (Paper 3) questions

3 (i) Compare the services available in Crank and Kings Moss. [2]

 (ii) Describe the general relationship shown by the table between the number of people who live in a settlement and the services available. [1]

 (iii) Suggest why settlements like Crank, Houghwood and Kings Moss do not have many services. [2]

Higher Tier (Paper 4) question

4 (a) To what extent is there a relationship between the population of these settlements and the types of services available ? You should refer to examples from the table in your answer.

[3]

Although the same concept – service provision in a rural area – is being examined in both papers, it is done in different ways. The questions on the Foundation Tier paper are more straightforward but they do have three different **commands**:

> compare
>
> describe
>
> suggest why.

You must understand what each command word means. Page 232 will help you to do this. The question on the Higher Tier paper has a smaller mark allocation and is more difficult. Notice the command words 'to what extent'. This instructs the candidate to make a judgement about the statement which follows and then support the judgement with examples taken from the table. Page 233 will help you to answer this type of question.

When this question was marked by examiners, the following rules were followed:

▶ 1 mark – recognise the relationship between population size of a settlement and the number of services

▶ 1 mark – recognise that there is an exception to this relationship and identify the exception (the secondary school)

▶ 1 mark – recognise that there is a relationship between the type or order of services available in different sized settlements.

The three rules show how some marks are easier to score than others.

The following questions concentrate on land use in a small area and use another type of resource – a large-scale street plan which shows more detail. The questions are taken from the Foundation Tier (Paper 1).

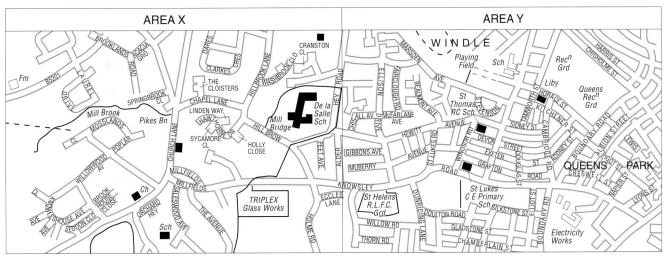

Edge of Town ◄──────────────────────────────────► CBD
(Central Business District)

(b) (i) Housing is the main land use in area Y. Use map evidence only to give one other land use in this area. [1]

The question focuses on area Y (not area X). The example must come from the map. It cannot be a general land use like shops.

(ii) Use map evidence only to describe one difference in the road layout between areas X and Y. [1]

This question again requires map evidence and the answer must show the difference between the two areas. Therefore you must use comparative words in your answer. See page 233 for advice on how to answer this type of question.

(iii) Give three likely differences between the houses in areas X and Y. [3]

Similarly this question is instructing the candidate to focus on differences, therefore the answer must be comparative. Unlike part (ii) the answer cannot be taken directly from the map so you must use your knowledge and your understanding of the concept that housing types will vary between different areas of a town.

The following question is similar in both Foundation and Higher Tier papers.

(c) The population of Rainford (in and around grid square 4801) has increased in recent years as people have moved from urban areas such as St Helens.

Foundation Tier question

5 Suggest any **three** attractions of living in Rainford. You should give evidence from the OS map extract in your answer. (3)

Higher Tier question

6 Use evidence from the OS map extract **only** to suggest the attractions of living in Rainford. (4)

The Higher Tier question instructs the candidate to use only map evidence while there is not the same restriction in the Foundation Tier question and therefore answers may be more general.

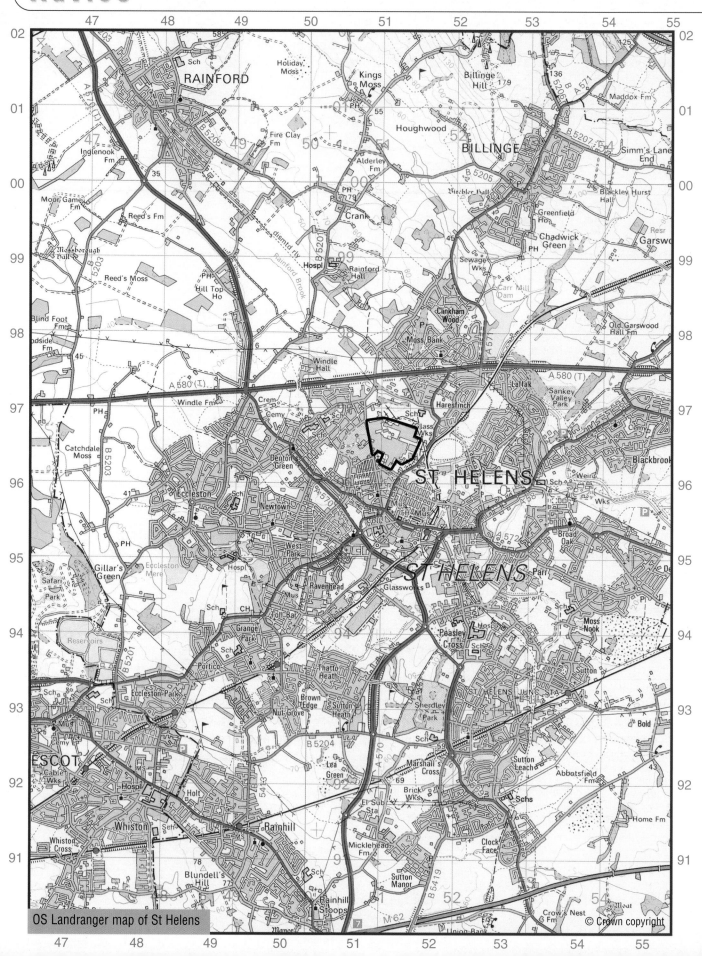

OS Landranger map of St Helens

© Crown copyright

ROADS AND PATHS   Not necessarily rights of way

	Motorway (dual carriageway)
M1	Motorway under construction
Unfenced Footbridge	Trunk road
A 470 (T) Dual carriageway	Main road
	Main road under construction
B 4518	Secondary road
A 855 Bridge B 885	Narrow road with passing places
	Road generally more than 4m wide
	Road generally less than 4m wide
	Path / Other road, drive or track
	Gradient: 20% (1 in 5) and steeper, 14% (1 in 7) to 20% (1 in 5)
	Gates / Road Tunnel
Ferry P Ferry V	Ferry (passenger) / Ferry (vehicle)

Service area — Junction number — Elevated

RAILWAYS

	Track multiple or single		Bridges / Footbridge
a	Station, (a) principal		Embankment
	Track narrow gauge		Cutting
	Freight line, siding or tramway		Tunnel
			Viaduct / Level crossing LC

WATER FEATURES

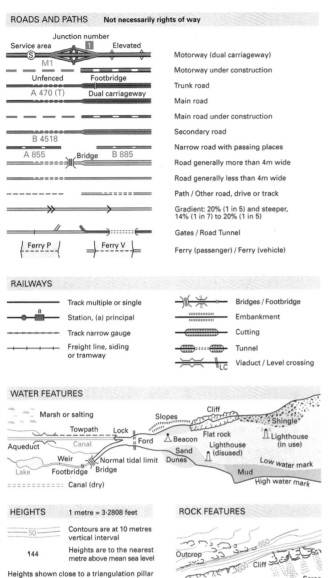

Marsh or salting
Towpath   Lock   Cliff   Slopes   Shingle
Aqueduct   Canal   Ford   Beacon   Flat rock
Weir   Lighthouse (disused)   Sand   Lighthouse (in use)
Lake   Footbridge   Bridge   Normal tidal limit   Dunes   Low water mark
Mud   High water mark
Canal (dry)

HEIGHTS   1 metre = 3·2808 feet

50 — Contours are at 10 metres vertical interval

144 — Heights are to the nearest metre above mean sea level

Heights shown close to a triangulation pillar refer to the ground at the base of the pillar and not necessarily to the summit.

ROCK FEATURES

Outcrop   650   Cliff   600   Scree

PUBLIC RIGHTS OF WAY

Not shown on maps of Scotland

- - - - - - -   Footpath
- · - · - · -   Road used as a public path
- - - - - - -   Bridleway
-+-+-+-+-   Byway open to all traffic

The symbols show the defined route so far as the scale of mapping will allow.

The representation on this map of any other road, track or path is no evidence of the existence of a right of way

Danger Area — Firing and Test Ranges in the area. Danger! Observe warning notices.

OTHER PUBLIC ACCESS

· · · ·   Other route with public access (not normally shown in urban areas). Alignments are based on the best information available. These routes are not shown on maps of Scotland.

● ●   National/Regional Cycle Network

—   Surfaced cycle route

4   National Cycle Network number

8   Regional Cycle Network number

◆ ◆   National Trail, Long Distance Route, selected Recreational Routes

ANTIQUITIES

+	Site of monument
· o	Stone monument
⚔	Battlefield (with date)
☆ ····	Visible earthwork
VILLA	Roman
Castle	Non-Roman

ABBREVIATIONS

CG	Coastguard
CH	Clubhouse
MP	Milepost
MS	Milestone
P	Post office
PC	Public convenience (in rural areas)
PH	Public house
TH	Town Hall, Guildhall or equivalent

LAND FEATURES

⊼—⊼—⊼	Electricity transmission line (pylons shown at standard spacing)
> - -> - ->	Pipe line (arrow indicates direction of flow)
ruin	Buildings
	Public building (selected)
	Bus or coach station
	Place of worship with tower
	Place of worship with spire, minaret or dome
+	Place of worship without such additions
o	Chimney or tower
	Glasshouse
Ⓗ	Heliport
△	Triangulation pillar
	Radio or TV mast
	Windpump / wind generator
	Windmill with or without sails
+	Graticule intersection at 5' intervals
	Quarry
	Spoil heap, refuse tip or dump
	Coniferous wood
	Non-coniferous wood
	Mixed wood
	Orchard
	Park or ornamental ground
	Forestry Commission access land
	National Trust-always open
	National Trust-limited access, observe local signs
	National Trust for Scotland

TOURIST INFORMATION

i i	Information centre, all year / seasonal
	Viewpoint
P	Parking
⊽	Picnic site
Å	Camp site
	Caravan site
	Selected places of tourist interest
✆ ✆	Telephone, public / motoring organisation
Γ	Golf course or links
▲	Youth hostel

BOUNDARIES

-+ - + -+	National
-·+·-·+·-	District
-··-··-··	County, Unitary Authority, Metropolitan District or London Borough
	National Park or Forest Park

The following answers show different levels of success in answering these questions.

> 1. *Would be away from CBD containing factories.*
>
> 2. *Short drive to the main shopping town.*
>
> 3. *Small village in the countryside.*

Examiner's comment

The candidate uses the OS map to give three reasons why Rainford is an attractive village in which to live. Although the three statements do not read well grammatically, they do show that the candidate realises that the rural area has advantages, especially for people with cars.

Candidate 2

> 1. *Rainford has a school and a church and good facilities.*
>
> 2. *It is less crowded around Rainford than in St Helens.*
>
> 3. *There are not many roads near Rainford and less noise and pollution.*

Examiner's comment

The candidate uses the OS map to identify services located in Rainford, but only one mark is allocated for these. The second point shows understanding of lower population density in the countryside. The third idea is almost acceptable but the link between noise and the bypass is not really made.

Candidate 3

> 1. *No noise from traffic.*
>
> 2. *There will not be much pollution in the area.*
>
> 3. *There will be a lot more space around.*

Examiner's comment

The candidate obviously has some understanding of why Rainford may be an attractive place to live but none of the ideas is sufficiently explained. The candidate needed to explain:

▶ why there is no (or less) traffic noise.

▶ the type of pollution and why it will be less.

▶ the type of open space and why there will be more.

Higher Tier answers
Candidate 1

In Rainford there is more open space. The area around it is very rural as it is on the rural/urban fringe. It is not very large and so would be peaceful, and as it has no CBD there will be no ugly office blocks and fewer traffic jams in congested roads leading to less air pollution. It is a nice scenic area with the river running through it and it is just off the road meaning that it is very quiet but also it is very easy to travel back into St Helens for a job. There is a school that you could send your children to and the houses are not terraced and have bigger gardens.

Examiner's comment

This is an excellent answer which shows full understanding of the question. The candidate supports the ideas by appropriate map evidence. The benefits of the rural environment are related to Rainford in a well-developed way.

Candidate 2

The attractions are:
there is room for expansion.
the land is flat.
it is made accessible by the A570 which however does not run through the town thereby reducing noise and pollution.
it has its own range of services such as a school and a church.
people could live here and travel to work in St Helens.
there are fields and farms nearby creating a pleasant atmosphere and environment.

Examiner's comment

This candidate also shows a good understanding of the attractions of the rural area. The points about room for expansion and flat land are not relevant, but these are followed by three well-developed ideas concerning noise pollution from traffic, local services, and the open rural environment. Unfortunately the idea about commuting to St Helens is not supported by map evidence.

Candidate 3

The housing is a lot spread out and the roads curve which means that there will probably be gardens and bigger homes. There are a few woods nearby where they can take their dogs for walks. There is only B and even smaller roads going through Rainford which means there will be less traffic.

Examiner's comment

The first idea is not supported by map evidence. The two ideas about the attractions of the local woods and less traffic are apparent from the OS map.

TASKS

1. The following question is based on the street plan of St Helens on page 121.

   Give three likely differences between the housing in areas X and Y.   [3]

   Now you become an examiner. Study the answers below to question 4 and decide how many marks each one scored in the examination.

Answer A
1. *Smaller houses and gardens in Y.*
2. *Larger houses and gardens in X.*
3. *Homes in Y are dearer because of being so close to town.*

Answer B
1. *Area X will have more detached houses than area Y.*
2. *The houses will be older in area Y.*
3. *They will also be more expensive in area X.*

Answer C
1. *The houses in this area will be older.*
2. *They will be closer together.*
3. *They will probably be terraced.*

Answer D
1. *Housing in area Y will be mostly terraced housing.*
2. *Housing in area X will be more expensive.*
3. *Housing in area X will be bigger than Y.*

Answer E
1. *There will be small, terraced houses in Y and semi-detached in X.*
2. *The roads will be smaller in Y.*
3. *The houses will be cheaper in Y.*

The actual marks scored were.
Answer A : 1   B : 3   C : 0   D : 2   E : 2

THE ENTRY LEVEL CERTIFICATE

Coursework tasks are a very important part of the Entry Level Certificate course. They are worth a total of 50% of the marks and you can work on the tasks at your own speed to improve your marks.

Coursework tasks can include some **data collection**. This means you can include some information that has been collected from the local area or from newspapers or surveys. You must also show that you can **present work** in a variety of ways. It could include maps, sketches, photos, tables of information or leaflets and newspaper cuttings which are about your work. The task should include a **commentary** where the work is described and explained in as much detail as possible.

Good examples of Coursework tasks can include a piece of **fieldwork**. It provides evidence that you have actually visited, observed and collected information. It does not have to be a long trip to collect the information. It could be collected in the place where you live.

People and Places to Live – a Coursework task

Settlements are places where people live, work and play. People live in many different types of settlements all over the world – **cities**, **towns**, **villages**, **hamlets** or in houses all on their own. The land inside settlements is used in many different ways, such as for shops, parks, factories and offices as well as houses.

In the UK, only 15% of the people live in **rural** areas – villages or hamlets in the countryside. 85% of the people live in **urban** areas – towns and cities. Here you will find a pattern for how the land is used. In the Central Business District (CBD) most of the buildings are shops and offices. There are very few places to live.

Most people live in the **inner city** or **suburbs**, in houses or flats or bungalows, away from the expensive centre of the settlement.

TASKS

Title – Places to Live

1 Make a copy of the table showing different sorts of buildings that people live in.

For each one, write a sentence to describe it. The first has been done for you.

Type of home	What it is like
Bungalow	House with no upstairs

2 Why do few people live in the CBD?

3 Draw a pie graph or a bar graph to show the percentage of people living in urban and rural areas of the UK.

4 Why do so many people live in urban areas in the UK? What are the attractions?

People live in many types of buildings in different parts of towns and cities. Photos A, B, and C show three different types of housing – terraced housing, detached and semi-detached .

In many towns, the terraced housing is the oldest housing. It was built in rows of streets near to the CBD. There will often be small corner shops, schools and old factories all close together in an area known as the inner city. Newer semi-detached and detached housing is usually found near the edge of towns in estates in the suburbs. On some estates, there are very few shops and services at all.

TASKS

5 Draw a plan or sketch of two of the houses in the photos. Label them with information about the features of the houses, such as building materials, rooms and size.

   * **Fieldwork Tip** – As part of your coursework, you could sketch your own house and one different type of house in your area. Label them with information.

6 Look at the **Heads and Tails**. Sort them out to match up.

Heads	Tails
Line of houses joined together	Flat
One house not joined to any other	Terraced
Two houses joined together	Detached
Several homes in one building	Semi detached

People sometimes move from one part of a town to another at different times in their lives. For example, an elderly couple may move from a semi-detached house to a bungalow when they retire. Once their children grow up and leave home, they need less space. As they get older, they may not be as able to climb stairs so easily.

TASKS

7 Explain why someone may move from a terraced house into a semi-detached or detached house, at some time in their life.

Some areas of the inner city are now over 100 years old. As a result, the houses may need to be repaired or modernised. Window frames may need replacing. The roof may need new tiles. Brickwork may need pointing. To make the houses more modern, some may need an inside toilet and bathroom or central heating adding. In some areas, roads have been blocked off and play areas and small garden areas made where the street used to be. The back alley ways may be paved and street lights fitted to make people feel safer.

In many cities, the terraced housing was demolished and replaced by multi-storey flats. The flats were sometimes over twenty stories high with balconies instead of back yards or gardens. The flats were fitted with lifts and parking and play areas were found at ground level.

TASKS

8 Think of different sorts of housing. Which would be best for:
  a an elderly couple in their 60s,
  b a middle aged couple with a young child,
  c a group of students at college?
  Write down why you chose each one.

9 a Which type of housing would you prefer to live in?
    b Write down your reasons.

■ People and their Needs

The quality of life

WHAT IS QUALITY OF LIFE?

Quality of life differences **FIGURE 3.1**

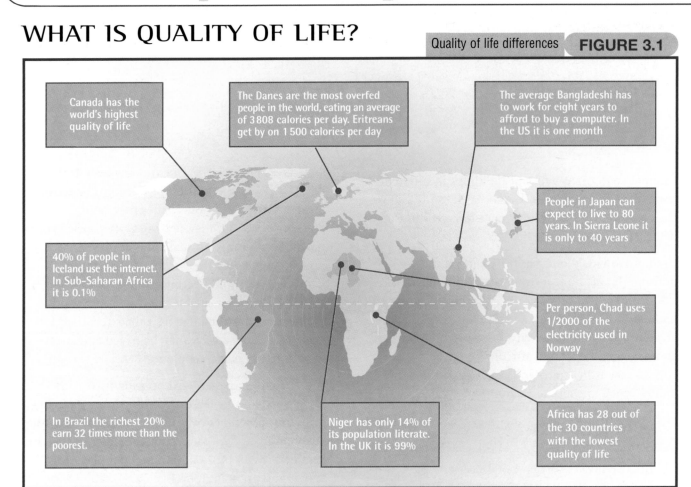

Canada has the world's highest quality of life

The Danes are the most overfed people in the world, eating an average of 3808 calories per day. Eritreans get by on 1500 calories per day

The average Bangladeshi has to work for eight years to afford to buy a computer. In the US it is one month

40% of people in Iceland use the internet. In Sub-Saharan Africa it is 0.1%

People in Japan can expect to live to 80 years. In Sierra Leone it is only to 40 years

Per person, Chad uses 1/2000 of the electricity used in Norway

In Brazil the richest 20% earn 32 times more than the poorest.

Niger has only 14% of its population literate. In the UK it is 99%

Africa has 28 out of the 30 countries with the lowest quality of life

Images on television help highlight the differences that exist between parts of the world. Contrasts in living conditions, the level of economic development, the health of people and difficulties of everyday life can be stark. The UK is described as a rich, industrialised nation that enjoys a **high quality of life**. We can compare our perception of the UK with other countries (Figure 3.1). What is your view of our level of development?

Figure 3.1 also shows some countries which have a higher quality of life than the UK. Identify them with the help of an atlas and consider whether any surprise you. Then do the same for some countries with a lower quality of life.

Indicators

Geographers use the **level of development** to describe the economic and social conditions of a country. They use statistical data or **indicators** to measure and compare countries.

One useful indicator is **Gross Domestic Product (GDP)**. This is a measure of the value of all the work done in a country. By comparing GDP, it is possible to identify the richest nation. However, it is likely that a country with a large population will produce more than a smaller one (Figure 3.2). It also depends on how developed the country's industry is. Therefore comparing the GDP per person for each country gives a more useful measure.

Country	GNP $millions	GNP/person $	Popn. millions
Argentina	280 000	8 100	35
Australia	338 000	18 700	19
Bolivia	6 000	800	8
Burkina Faso	2 500	230	11
China	750 000	620	1 210 000
Egypt	45 600	790	63
France	1 451 000	25 000	58
India	319 000	340	980 000
Portugal	96 700	9 800	10
UK	1 095 000	18 700	58 600
USA	7 100 000	27 000	268 000
Zambia	3 600	400	9 500

Figure 3.2 GNP and population data for selected countries

Using just one indicator can have some disadvantages. For example, GDP can be misleading because it is an average. Wealth is not evenly spread across the population.

Profiles and indexes

Using several different indicators provides a broader basis to describe and compare countries Information can then be used to write a profile of a country.

The United Nations uses an index called the Quality of Life Index. It uses **life expectancy**, **infant mortality** and **adult literacy** to measure the quality of life (Figure 3.3). The World Development Report ranks countries using the purchasing power of the local currency.

TASKS

1 What is meant by the term 'Quality of Life'?
2 Suggest five different indicators. For each one, describe what it measures.
3 Explain why using a single indicator is likely to be unreliable.
4 a Use the information in Figures 3.1 and 3.2 to write a description of the quality of life in the UK.
  b Choose an LEDC and use the information in Figures 3.1 to 3.3 to write a description of that country.

FIGURE 3.3
Quality of Life Index

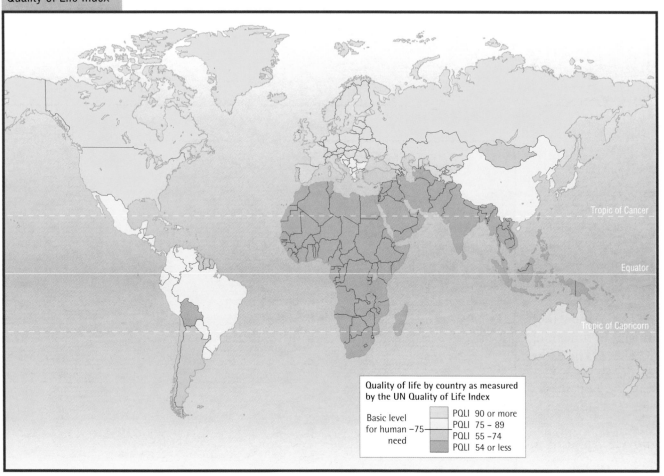

Quality of life by country as measured by the UN Quality of Life Index

Basic level for human need —75—

PQLI 90 or more
PQLI 75 – 89
PQLI 55 –74
PQLI 54 or less

Measuring development

Quality of Life indicators show a global pattern of development. They reveal a North–South divide (Figure 3.4). North of the line are the more economically developed countries (MEDCs). To the south are the less economically developed countries (LEDCs). Note which side of the divide Australia and New Zealand are.

We can compare indicators to see if a change in one causes a change in another. A hypothesis may then be used to investigate them, for example, the higher the GDP per person, the longer the life expectancy. We can then say that here is a **correlation** between the two indicators.

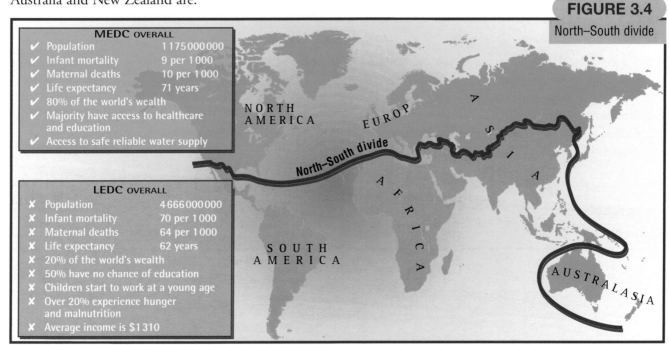

FIGURE 3.4
North–South divide

MEDC OVERALL
- ✔ Population — 1 175 000 000
- ✔ Infant mortality — 9 per 1000
- ✔ Maternal deaths — 10 per 1000
- ✔ Life expectancy — 71 years
- ✔ 80% of the world's wealth
- ✔ Majority have access to healthcare and education
- ✔ Access to safe reliable water supply

LEDC OVERALL
- ✘ Population — 4 666 000 000
- ✘ Infant mortality — 70 per 1000
- ✘ Maternal deaths — 64 per 1000
- ✘ Life expectancy — 62 years
- ✘ 20% of the world's wealth
- ✘ 50% have no chance of education
- ✘ Children start to work at a young age
- ✘ Over 20% experience hunger and malnutrition
- ✘ Average income is $1310

NORTH AMERICA — EUROPE — ASIA — AFRICA — SOUTH AMERICA — AUSTRALASIA

North–South divide

It is sometimes possible to see a relationship by mapping indicators with graded shading. These maps use a range of colours, a single colour or shading to show the amount of something in an area. The darker the shading, the higher the value (Figure 3.5). By comparing maps, it possible to identify similar patterns.

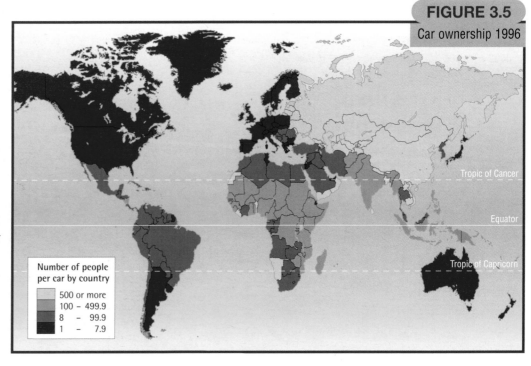

FIGURE 3.5
Car ownership 1996

Number of people per car by country
- 500 or more
- 100 – 499.9
- 8   – 99.9
- 1   –   7.9

Tropic of Cancer

Equator

Tropic of Capricorn

A scattergraph plots two variables and indicates whether there is a link between the data. This is shown by drawing on a line of 'best fit'. This is a straight line that comes close to or joins many of the points on the graph (Figure 3.6).

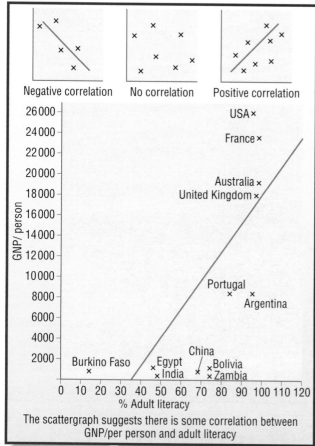

The scattergraph suggests there is some correlation between GNP/per person and adult literacy

| FIGURE 3.6 | Scattergraph of adult literacy against GNP |

To see if there is a correlation, we can use a statistical method called Spearman Rank Correlation. This is outlined as follows:

$\Sigma d^2 = 49$

$r^s = 1 - \left(\dfrac{6 \times 49}{12^3 - 12} \right)$

$r^s = 1 - \left(\dfrac{294}{1716} \right)$

$r^s = 1 - 0.171$

$r^s = \quad 0.8$

This suggests that a good correlation exists between GNP/person and adult literacy. Can you suggest a reason why?

▶ Choose two sets of data, for example, GDP per person and adult literacy.

▶ Rank each set of data. The highest is Rank 1.

▶ Subtract the smaller rank number for each country line from the larger. For example, USA is rank 1

for GDP person but rank 3 for adult literacy. Put the answer 2 in the Difference (d) column.

▶ Multiply the difference by itself. For example, USA's difference of 3 squared is 9. Write this in the Difference Squared (d²) column.

▶ Add up all the figures in the Difference Squared column.

▶ Now substitute the figures into the following formula:

$$r^s = 1 - \left(\dfrac{6 \, \Sigma \, d^2}{n^3 - n} \right)$$

Where: d is the difference between the ranks,
Σ is the total or sum of,
n is the number of countries
(there are ten so $n^3 - n$ will be 1000 − 10 = 990).
r^s is the answer, or the Spearmans Rank Correlation Coefficient. The nearer the answer is to 1.0 the better the correlation, at 0.0 there is no correlation. 0.7 upwards is a good correlation.

Country	Rank GNP/person	Rank adult literacy	d	d²
Argentina	6	5	1	1
Australia	3	1=	2	4
Bolivia	7	8	1	1
Burkina Faso	12	12	0	0
China	9	6	3	9
Egypt	8	11	3	9
France	2	3	1	1
India	11	10	1	1
Portugal	5	7	2	4
UK	4	1=	3	9
USA	1	4	3	9
Zambia	10	9	1	1

Figure 3.7 Rank Spearman Correlation

TASKS

1 Using Figure 3.4:
   a Which parts of the world have a high quality of life?
   b Which areas of the world have a low quality of life?
   c Describe the pattern shown.
2 Suggest the link between GDP per person and infant mortality. How might it be explained?
3 Using the data from Figure 3.2, investigate the following hypotheses using one or more of the methods and suggest reasons for any correlation.
   a Countries with a high GDP per person also have a long life expectancy measure.
   b A low quality of life and short life expectancy are related indicators.
   c There is a link between the number of doctors and GDP per person.

COMPARING THE UK WITH BURKINA FASO

Burkina Faso is a country of 9.8 million people in Central Africa. It is a landlocked, dry, **savanna** environment (Figure 3.8).

Burkina Faso is a country 'locked in deprivation' (Figure 3.9). It is not quite the poorest country in the world. Its average income of US$230 per year is higher than 13 others. But world economic changes beyond the control of its government have reduced the purchasing power of its currency in recent years.

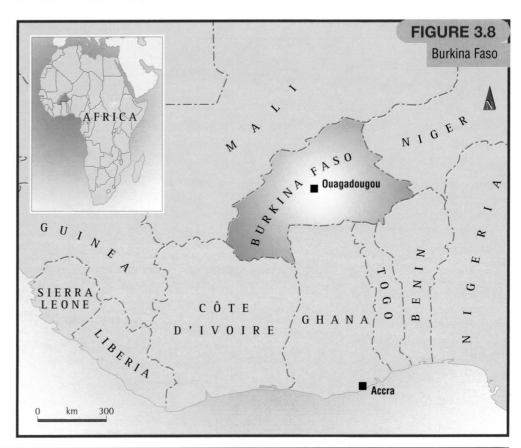

FIGURE 3.8
Burkina Faso

Looking for a road out of nowhere

ECONOMIC commentators rarely report from Burkina Faso. When they do travel, it is usually to the capitals of the developed countries, the rich world, or perhaps to the quickly becoming rich world – the booming economies of East Asia

Reporting on the very poor countries of sub-Saharan Africa is left to other journalists: writers who specialise in the problems of economic development, or those who report on famines or coups. This seems a pity, most obviously because, although sub-Saharan Africa may not matter much in world economic terms, it matters enormously in political and human terms.

But there is another powerful reason why more should be written

about these countries: this is that quite small changes in the economic behaviour of rich countries can have a disproportionate impact on the poor, and our policy makers should be reminded of that.

But it does not feel like a disaster zone; just that life for most people is very precarious. That may be something to do with the good rains last summer. Or it may be that poverty in the countryside (and Burkina Faso is exceptionally rural, with only 18 per cent of its population in towns) always seems less gnawing than urban poverty.

What does feel quite extraordinary – and gives a measure of poverty – is the lack of cars. An apparently prosperous

town of 2,000 or 3,000 people will have not a single private car. There is hardly a petrol pump between Ouagadougou and Po, 100 miles to the south on the Ghana border, and not a working one in Po itself

The area is not a disaster zone – yet. The population will double in the next generation, raising the brutal question as to how this country can manage to feed, clothe and house these extra people. It cannot hope to employ them: already an estimated one million Burkinabes work abroad, mostly in the neighbouring Ivory Coast.

INDEPENDENT ON SUNDAY
19 February 1995

Hamish McRae

Figure 3.9

Health	
Infant mortality	6 per 1000 births
Malnourishment	2% of the population
Average calorie intake per day	3317 calories
Maternal deaths	7 per 1000 births
Fertility rate	1.7 children per women
Life expectancy	77 years
Birth rate	13 per 1000
Death rate	11 per 1000
Population doubling time	433 years
Access to healthcare	100%
Population per doctor	300 people
Social	
Adult literacy	99%
GDP spent on education	5.3%
Urban population	90%
Access to safe, reliable water	98%
Average income per year	$18 000

Figure 3.10 UK quality of life indicators

The quality of life indicators compare starkly with those for the UK (Figure 3.10). Children, the most vulnerable group have a difficult life. Mortality is high among under fives: 17% of children die before the age of five and 30% of children are malnourished. The average calorie intake is below the minimum needed for a healthy diet. A tenth of mothers die during or as a result of childbirth. Mothers have on average 6.9 children each. Life expectancy is 48 years. Educational levels are exceptionally low, adult literacy is 18%, with only 36% of boys and 23% of girls attending school. The 3% of GDP spent on education has little impact. With a high birth rate and a population projected to double in the next 23 years, it will be difficult for Burkina Faso even to maintain its present low quality of life.

Countries like Burkina Faso seldom make the news. Yet life for its population is very difficult. The unreliable rains threaten agricultural production. Clean and reliable water is available to only 56% of the population and less than half of the population has access to basic healthcare. It is a rural nation with only 18% living in urban areas. Rural poverty poses issues different from urban lifestyles.

FIGURE 3.11

Market day, Burkina Faso

TASKS

1 Describe the location of Burkina Faso.

2 Explain why life in Burkina Faso is described as precarious in Figure 3.9.

3 Choose five quality of life indicators and compare Burkina Faso with the UK.

4 How well does Burkina Faso meet these basic needs of its people:
   a access to healthcare
   b access to clean water
   c access to education.

5 What problems and issues will Burkina Faso need to address in the next 20 years?

6 How is poverty in the countryside different from poverty in the towns of Burkina Faso?

7 Make a more detailed comparison of Burkina Faso and the UK by using a CD-ROM atlas or by downloading relevant data from Internet sites such as The CIA World Fact Book:

http://www.odci.gov/cia/publications/factbook/

ECONOMIC INDICATORS OF DEVELOPMENT

The employment structure of a country is another way of showing how developed a country is. Industry can be classified as **primary**, **secondary** or **tertiary**.

▶ **Primary industry.** This provides the raw materials used by people and industry, for example through farming, mining, fishing and forestry. Unprocessed goods are often of low value and include items such as hardwood, mineral ores, and agricultural products such as cotton.

▶ **Secondary (or manufacturing).** These industries use raw materials to make other products. This adds value to the raw materials. For example, iron ore is used to make steel for cars and wood is used to make furniture.

▶ **Tertiary (or services).** These industries do not produce raw materials or manufactured goods. Instead, they provide specialist services such as banking, administration, healthcare, education, transport and entertainment.

As a country becomes more developed the proportion of people employed in each industrial sector changes. Figure 3.12 shows this for USA and Egypt, and Figure 3.13 shows recent changes in the UK.

	USA			Egypt		
	primary	secondary	tertiary	primary	secondary	tertiary
1900	38	27	35	71	10	19
1920	28	33	39	65	11	22
1940	19	35	46	70	10	20
1960	7	39	54	57	12	31
1980	4	35	61	50	14	36
2000	3	25	72	42	21	37

Figure 3.12 Changes in occupational structure USA and Egypt

Figure 3.14 shows the proportions employed in each sector for eleven different countries. As you read down the list, the number in primary industry declines and the numbers employed in secondary and tertiary increase. These differences can be shown using a triangular graph (Figure 3.15).

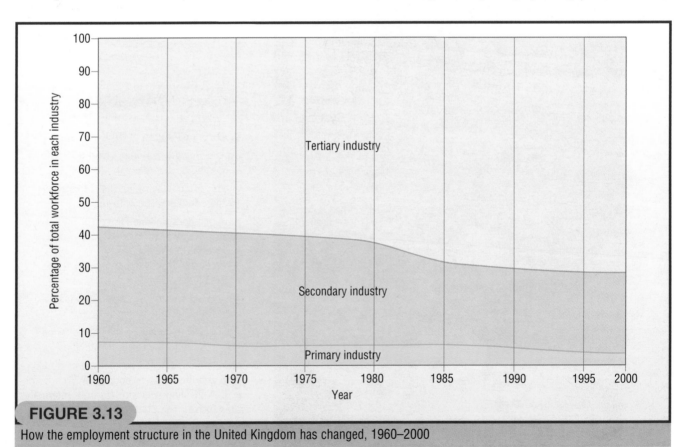

FIGURE 3.13

How the employment structure in the United Kingdom has changed, 1960–2000

Country	Agriculture	Industry	Services	Pop. growth	Income
Bangladesh	59	13	28	1.0	$240
Brazil	25	25	50	0.8	$3640
Chad	83	5	12	2.5	$180
Egypt	42	21	37	2.6	$790
France	9	29	62	0.6	$20 580
Ghana	59	11	30	2.7	$390
India	62	11	27	2.5	$340
Italy	9	32	59	0.2	$19 020
Japan	7	34	59	0.3	$39 000
Sweden	3	28	69	0.7	$23 750
UK	2	28	70	0.3	$18 700

Figure 3.14 Occupational structure for selected countries

Other economic indicators such as energy use per person, origin of income, **infrastructure** and trade per person can also be used to describe the development of a country. In some LEDCs, however, industrial development is hampered by the rate of population increase. There are not enough jobs in the secondary or tertiary sectors so more people have to rely on the primary sector to make a living.

TASKS

1 a Do a survey among your teaching group. List and categorise the range of jobs found in students' families.

b Copy the table below to list your results.

Primary	Secondary	Tertiary

c Present your results as a bar chart.
d Describe and suggest reasons for your results.

2 Describe what you would find if you undertook a similar survey in an LEDC. Suggest reasons why the results would be different.

3 Describe how the employment structure is different between an LEDC and an MEDC.

4 Explain why there is a link between GDP per person and employment structure.

5 Use the data in Figure 3.14 to plot three scattergraphs. Write a description of the link between the economic indicators.

a energy used per person and percentage employed in secondary industry.
b GDP per person and percentage employed in primary.
c one of your own choice.

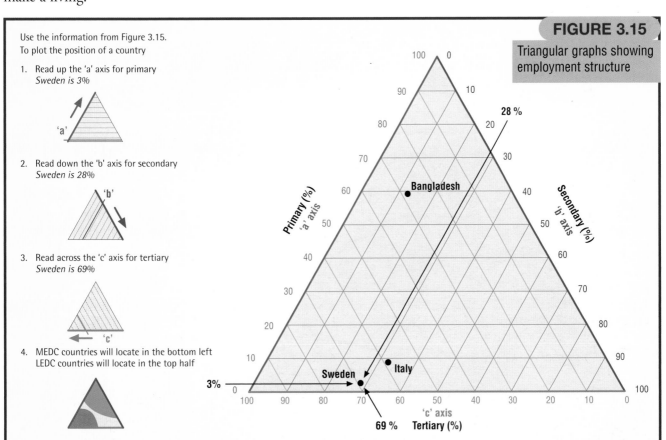

FIGURE 3.15

Triangular graphs showing employment structure

Use the information from Figure 3.15.
To plot the position of a country

1. Read up the 'a' axis for primary
   Sweden is 3%

2. Read down the 'b' axis for secondary
   Sweden is 28%

3. Read across the 'c' axis for tertiary
   Sweden is 69%

4. MEDC countries will locate in the bottom left
   LEDC countries will locate in the top half

INDUSTRIALISATION AND ECONOMIC GROWTH

Different countries are at different stages of development. The UK experienced rapid industrialisation in the nineteenth century, together with many other European countries and the USA, and developed a wide range of industries. These countries traded with their colonies to gain raw materials and sell manufactured goods. This created wealth and improved the quality of life of their people.

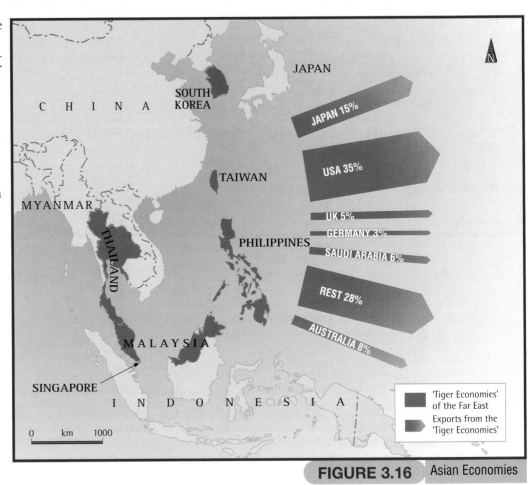

FIGURE 3.16 Asian Economies

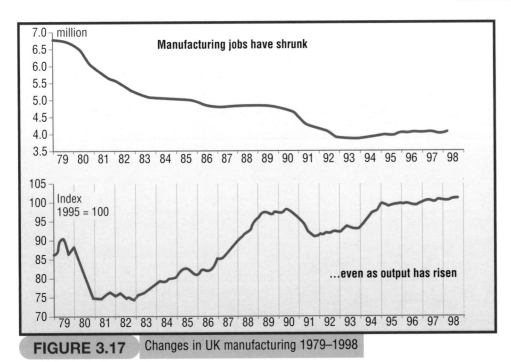

FIGURE 3.17 Changes in UK manufacturing 1979–1998

In Asia, Japan was first to industrialise. During the twentieth century, other regions emerged as industrial nations capable of world trade: Hong Kong (now part of China), Taiwan, South Korea and Singapore. These countries are called **newly industrialised countries, (NICs)**. The Asian economies grew rapidly and became known as 'Tiger Economies' (Figure 3.16).

Industrialised countries are those which derive at least 25% of their GDP from industry, with most from manufacturing, and have about a third of their workforce employed in secondary industry. The growth of NICs has meant that the world's leading producers of manufactured goods have faced increasing competition. The USA's share of world manufacturing has almost halved in the last 20 years. The UK experienced a similar fall from 6.5% to 3.3%. This has meant a change in the UK's employment structure (Figure 3.17). Japan has 13.7% of world manufacturing trade. In contrast, the South Asian NICs account for 15% of total manufacturing.

An industrialising country passes through a series of stages. At each stage there are distinctive changes to the country's infrastructure and production (Figure 3.18). However, LEDCs do not have the economic or technological advantages of MEDCs. MEDCs can influence world markets and exploit resources. This can hamper the development of LEDCs. **Trans-national companies** (**TNCs**) which are based in MEDCs and operate worldwide, are a dominant force controlling about 25% of world trade. TNCs provide much needed investment for an LEDC but can prevent the LEDC developing its own industries, necessary to build a manufacturing base for its own development.

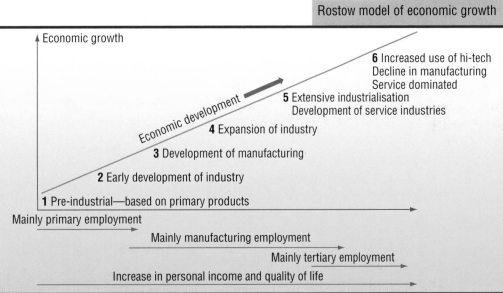

FIGURE 3.18

Rostow model of economic growth

The success of the Tiger Economies has been built on their governments' ability to control the industrialisation process. The initial investment was used to generate money for further investment and attract foreign money. This spiral of growth is called the **multiplier effect** (Figure 3.19).

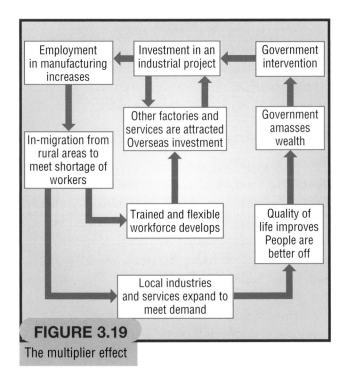

FIGURE 3.19
The multiplier effect

TASKS

1 Which countries are known as 'Tiger Economies'?

2 Write down one advantage and one disadvantage to LEDCs of trans-national companies.

3 Explain how the multiplier effect works.

SOUTH KOREA

None of the Tiger Economies has any extensive natural resources. South Korea is a mountainous peninsula with limited coal and iron deposits, but its real resource is its cheap and flexible workforce. From the late 1950s, the government backed investment in large companies capable of earning foreign currency. This was then invested in research and development projects and state-owned works like the Pohang Iron and Steel Corporation (POSCO). South Korea also has companies, such as Daewoo, Samsung, Goldstar, Kia and Hyundai, which are powerful family businesses. These were aided by cheap loans and high import taxes that made foreign goods more expensive than those made in Korea. The government kept living costs low, which helped to keep wage levels low, but this also depressed rural incomes. Between 1967 and 1976, 7 million rural workers migrated to the cities. This created a pool of cheap labour that helped the rapid expansion of the textiles, shipbuilding, vehicles and electronic industries (Figure 3.20). This made Korean exports cheap and gave South Korea a very high average annual growth rate of 7% between 1985 and 1990. In 1998, however, this fell back to 1%.

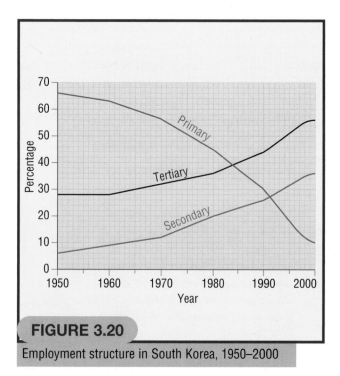

FIGURE 3.20

Employment structure in South Korea, 1950–2000

Steel industry

The state-owned POSCO has increased its production from 1 million tonnes in 1971 to 50 million tonnes in 2000. It imports coal, oil and iron ore from Australia, USA and Canada. Smaller companies produce specialist steels, but the steel industry is controlled and protected by the government and is regarded as an important foreign currency earner.

Shipbuilding

The growth in the oil trade in the 1960s made South Korea the leading producer of bulk carriers within 20 years. It now has a third of the world market in shipbuilding. In 1993, companies such as Samsung and Hyundai reduced their prices to capture more orders over its rival, Japan, for replacing ships built during the 1970s. With South Korea's lower labour costs and modern yards, they were able to win many more orders. However, South Korea's industry is now facing competition from Chinese yards.

Cars

The South Korean car industry is only 30 years old. Hyundai, Kia and Daewoo set up links with Japanese and American companies to develop the industry. There is slow growth in the domestic market so most of the cars are exported, some as kits for assembly elsewhere, using local components. Korean manufacturers are also involved in investing in LEDCs to exploit emerging markets.

Investment in South Korea

TNCs that invested in South Korea were attracted by the low production costs, lack of organised trades unions and closeness of the large Chinese market. Japanese TNCs, like Sony, built semi-conductor and television factories, taking advantage of special export production zones. By 1990, South Korean companies had closed the high technology gap between themselves and Japan and the USA. By 1992, South Korean TNCs, like Samsung, were investing in electronics factories in the USA and Europe (Figure 3.21).

Problems caused by rapid industrialisation

South Korea's rapid growth has had some negative effects. These include:

▶ **Social problems.** There are huge differences in the wages paid to women and immigrant workers when compared to men. Women are paid 75% and immigrants only 50% of male wages. Concern over working conditions and other restrictions have caused unrest.

▶ **Environmental problems.** The rapid industrial growth has polluted the environment and destroyed habitats. The increase in car ownership has prompted the capital, Seoul, to impose taxes on parking and road use, and offer incentives for people to change from coal to gas.

▶ **Lack of investment in research and development.** As world trade becomes more competitive, trade agreements which shared technology between South Korea and Japan have decreased. This has revealed South Korea's lack of research projects.

▶ **World trade.** In 1997, this faltered leaving many South Korean banks with huge debts or even bankrupt. The South Korean currency was devalued and this made Korean goods competitive for a while. But this also meant that South Korea reduced its imports from many other Asian countries.

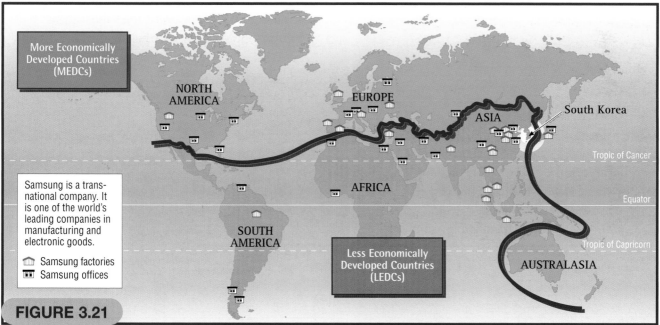

FIGURE 3.21

Samsung globalisation

	1970	1995	
	Number produced	Number produced	World share 1995
Televisions	114 000	17 102 000	12.7% 2nd
Tyres	900 000	53 472 000	5.9% 5th
Ships	0	5,000	33% 2nd
Cars	13 000	2 000 000	5.6% 5th
Commercial vehicles	15 000	510 000	3.3% 5th
Steel	480 000 tonnes	36 million tonnes	4.9% 6th
Energy use	1.37 tonnes per person	3.77 tonnes per person	
GNP per person	$4000	$9700	

Figure 3.22 South Korea: economic growth

TASKS

1 Describe the location of South Korea.

2 Explain how each of these factors helped South Korean companies develop:
   a cheap, flexible workforce,
   b government support with high import taxes,
   c low cost of living for the workers.

3 Give three examples of South Korean companies. List some of their products you can buy in the UK.

4 Using Figure 3.18, describe the economic growth of South Korea.

5 Why are South Korean TNCs investing in other countries?

6 Describe the costs and benefits of South Korea's economic growth.

FARMING SYSTEMS

There are many different ways of classifying farming activities. In MEDCs like the UK, farmers grow crops and rear animals for sale, hoping to make a profit. This is called **commercial farming**. In LEDCs there is also commercial farming but many farmers grow only enough to feed their families. This is called **subsistence farming**. Farming can also be classified according to:

Specialisation

▶ An **arable** farm grows crops
▶ a **pastoral** farm rears animals
▶ a **mixed** farm grows crops and rears animals.

Intensity of land use

▶ **Extensive farming** is where the farm size is very large compared with either the amount of money spent on it or the number of people working there.
▶ **Intensive farming** is where the farm is small in size compared to the numbers working there or the amount of money spent on it.

Nomadic farming is where farmers move from one area to another. Most farming is 'sedentary', that means it is based in the same place every year.

Farming operates as a system, with a series of inputs, processes and outputs. Inputs are what go into a system: they can be divided into physical, human and economic inputs. Processes are the activities on the farm which turn the inputs into outputs. Outputs are the products of the farm. If the farm is to make a profit the value of the outputs should be greater than that of the inputs.

TASKS

1 Explain the difference between:
   a commercial and subsistence
   b arable, pastoral and mixed farming
   c intensive and extensive farming
   d nomadic and sedentary farming.

FIGURE 3.23 Farm systems

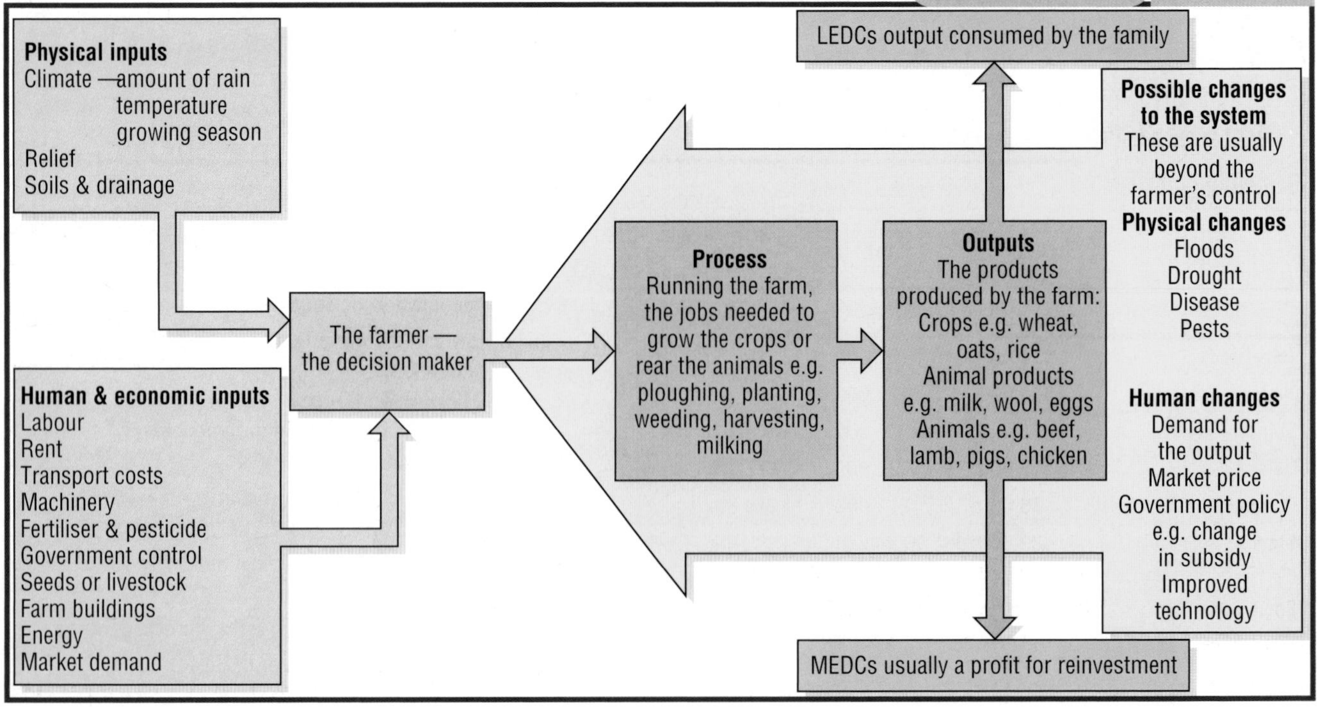

Physical inputs
Climate —amount of rain
         temperature
         growing season
Relief
Soils & drainage

Human & economic inputs
Labour
Rent
Transport costs
Machinery
Fertiliser & pesticide
Government control
Seeds or livestock
Farm buildings
Energy
Market demand

The farmer —
the decision maker

Process
Running the farm, the jobs needed to grow the crops or rear the animals e.g. ploughing, planting, weeding, harvesting, milking

Outputs
The products produced by the farm:
Crops e.g. wheat, oats, rice
Animal products e.g. milk, wool, eggs
Animals e.g. beef, lamb, pigs, chicken

LEDCs output consumed by the family

Possible changes to the system
These are usually beyond the farmer's control
Physical changes
Floods
Drought
Disease
Pests

Human changes
Demand for the output
Market price
Government policy e.g. change in subsidy
Improved technology

MEDCs usually a profit for reinvestment

TASKS

2 Farming systems often vary within countries and between countries, because the inputs are different. As a result of the decisions made by a farmer, the processes and outputs will be different. Make copies of the diagram below and complete them to show the following farming systems:

a subsistence farm in an LEDC

b arable farm in an MEDC.

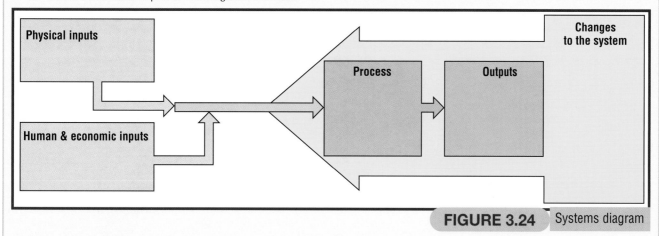

FIGURE 3.24 Systems diagram

If you have access to the Internet, the National Farmers Union website http://www.nfu.org.uk/ has an excellent education section which includes farm case studies from around the world. You could use these resources to complete Task 2 or you could produce the system diagrams with the aid of ICT. Using a desktop publishing program, copy and paste data from the NFU website into your diagrams.

A COMMERCIAL FARM IN AN MEDC

FIGURE 3.25

Thorn Park Farm

Dairy farming at Thorn Park Farm, North Yorkshire

Thorn Park is a dairy farm to the west of Scarborough. It is located in an undulating valley floor, surrounded by hills, which provide shelter from the prevailing winds which blow off the North Sea. The farm is run by the Wilson family, who rent the land from Scarborough Borough Council. Much of the work on the farm is done by Peter and his father Chris, but both of their wives also help out, particularly at milking times. The main part of the farm is 72 ha in size, although a further 36 ha are rented at Irton, 8 km away. They have a herd of 110 Friesian dairy cows. Each cow produces an average of 7 000 litres of milk a year. As well as the members of the family, the farm has one full-time employee.

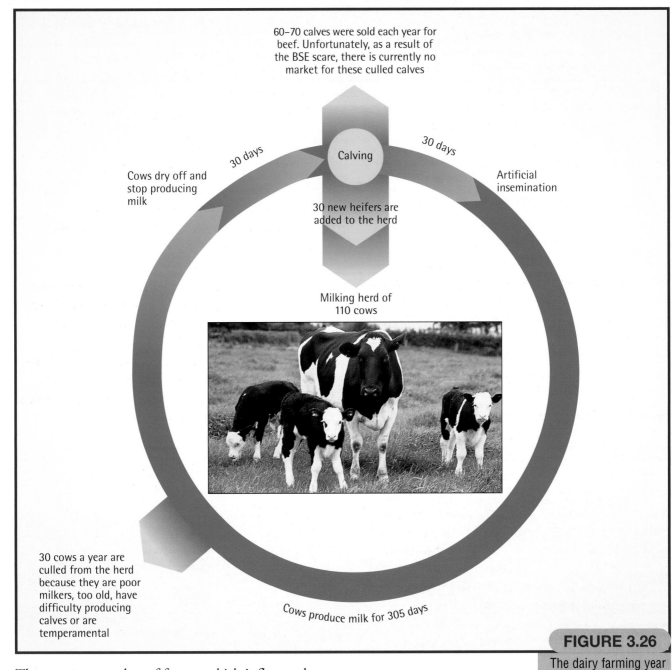

60–70 calves were sold each year for beef. Unfortunately, as a result of the BSE scare, there is currently no market for these culled calves

30 days

Calving

30 days

Cows dry off and stop producing milk

Artificial insemination

30 new heifers are added to the herd

Milking herd of 110 cows

30 cows a year are culled from the herd because they are poor milkers, too old, have difficulty producing calves or are temperamental

Cows produce milk for 305 days

FIGURE 3.26

The dairy farming year

There were a number of factors which influenced the Wilsons in their decision to be dairy farmers. The climate, although mild, is too wet, and the soils too heavy, to grow most arable crops. Also, the farm is too small to grow crops profitably. When Chris Wilson began farming the area, dairying was the best way to earn a living from the land. This is a good grass growing area.

Running a dairy farm requires full-time commitment: the herd of cows need to be milked twice a day, at 0600 and 1700 hours, *every* day of the year. Each milking session takes between one and two hours. The milk is stored in two large refrigeration tanks and is collected by bulk tanker every morning. There are plenty of other jobs to be done on the farm during the rest of the day. From April to November the cattle graze in the fields. Some of the fields are used to grow grass to be cut for silage. This is cut at least twice a year and stored around the farm buildings to be used as winter feed. It is important that the farmer gets sufficient cuts of grass to last the winter, otherwise it will be necessary to buy in feed, which will reduce profits. In late October and November, as the grass in the fields loses quality, the cows strip-graze a field of kale, a kind of cabbage. This supplements their diet and maintains milk quality. During the winter, the cows are kept in barns and fed on silage. Throughout the year, the farmer has to maintain the health of the herd and keep detailed records of each cow. The main sources of income on the farm are milk and calves. One of the major by-products of the farm is manure. This is collected and spread back onto fields, acting as a natural fertiliser.

A number of changes in the last few years have had serious implications for the success of the farm. The introduction of milk quotas by the European Union reduced the farmer's earnings. Milk quotas were introduced in 1984 to cut over-production in Europe. Farms were allowed to produce 90% of the production of the previous year. So, a dairy farm which produced 300 000 litres of milk from April 83 to March 84 was granted a milk quota of 270 000 litres. There have since been further reductions in quota of around 9%.

The greatest challenges, however, have occurred more recently. A disease called BSE has meant that the price of beef has fallen dramatically. 9% of the Wilsons' output at Thorn Park Farm used to come from the sale of calves, a further 10% from cows culled from the herd. The price for calves has fallen by 90% in three years, the price for culled cows by 50%. If this were not bad enough, the price Chris Wilson receives for his milk has fallen by 37%. At the same time, the costs of inputs on the farm have

either gone up or remained the same. The immediate future looks bleak for Thorn Park Farm. The Wilsons have recently been exploring their options. They could:

▶ Sell up and leave farming, but they would get very little from selling their livestock, and the land is only rented from the local council.

▶ Diversify into non-farming activities. The Wilsons have investigated setting up a caravan site on the land. They have the advantage of being in the North York Moors National Park next to Forge Valley, a honeypot site, as well as being close to Scarborough, a tourist resort. Unfortunately, being in a national park also means there are strict controls stopping developments like caravan sites.

▶ Try to last out the crisis, while other dairy farmers become bankrupt, hoping that eventually there will be a shortage of milk, forcing up prices. During this time the Wilsons would buy more milk quota and gradually increase the size of the milking herd up to 160 cows. By making farm processes more efficient to cope with the increase, they should eventually make a profit.

They chose the third option.

TASKS

1 Look carefully at the Ordnance Survey map on pages 44–45. Use it to draw a sketch map to show the site and situation of Thorn Park Farm. Label key features on your map such as the boundaries (grid reference 98 3882) of the farm, the surrounding hills, and Scarborough.

2 Draw a sketch cross section across the valley and farm from 982893 (Suffield Ings) to 987870 (Osborne Lodge), and label the position of the farm and other features.

3 Explain why the Wilsons decided to become dairy farmers.

4 Draw a systems diagram for Thorn Park Farm.

5 Explain the changes that made the Wilsons think about giving up farming.

6 Find Betton farm on the Ordnance Survey map on pages 44–45. It is located at 000854.

   This farm has also changed in recent years. Unlike Thorn Park Farm it has diversified. It now has a farm shop, a honey bee museum, an adventure playground, a nature trail, a restaurant, a collection of shops and craft centres, and animals for visitors to see. It has its own tourist brochure, and includes promotional material in the Scarborough Tourist guide.

   Look carefully at the OS map, comparing the two farms. Identify map evidence to explain why it has been possible for Betton farm to diversify, but not Thorn Park Farm.

FIGURE 3.27

Cutting grass for silage on Thorn Park Farm

SUBSISTENCE RICE FARMING IN INDIA, AN LEDC

The importance of rice

There are several thousand varieties of rice. Almost all require very moist soils throughout most of their growing season. Rice is a major world crop, feeding one third of the world's population. Its growing requirements mean that it can only be successfully cultivated in a few areas of the world. The main rice areas are in South East Asia. Rice growing is particularly suited to a monsoon climate as shown in Figure 3.28. Much of the rice grown in these countries is for subsistence – to feed the farmer and his family. Many farms are very small. They may have an area of only one hectare (the size of a football pitch) and be divided into as many as 15 plots. The small size of the farms together with the poverty of the people means that there is little mechanisation. The farming is labour intensive with much human effort required to do all the jobs. Water buffalo (oxen) are used to prepare the padi-fields and also provide manure, which is used as a means of supplementing soil fertility.

India is the world's second largest producer of rice, growing 20% of the world's total. Rice is the staple food of 65% of the total population in India, forming 90% of the country's total diet. Agriculture is the backbone of the country's economy, providing direct employment to about 70% of working people in India.

The Green Revolution is the name used for changes in farming in LEDCs such as India in the last 50 years. In the 1960s there was a fear that it would not be possible to grow enough food for the rapid increase in the world's population. It led to a search for new ways to increase agricultural productivity. In 1959 the International Rice Research Institute (IRRI) was set up in the Philippines to discover how rice yields could be increased.

• High yield varieties

MEDCs such as UK, USA, Germany and Australia provided money to develop high yield varieties (HYVs) of rice as well as wheat and maize. In 1965 an improved strain of rice called IR-8 led to an increase of rice yields of 300%. However, the increase in yields led to a fall in prices. Faster growing varieties have allowed an extra crop to be grown each year. Yields have also become more reliable as many of the new varieties are resistant to disease. HYV crops have shorter and stiffer stems, making them more resistant to wind and rain. However, HYVs require large amounts of fertilisers and insecticides as they are prone to insect attack. The environmental impact of increasing use of chemicals has meant that farming has become less sustainable as a result of soil erosion. Farmers who could afford to buy HYVs and fertiliser have benefited, greatly increasing their yields.

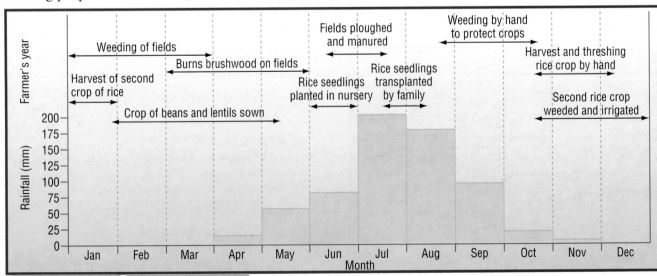

FIGURE 3.28 The rice farmer's year

THE GREEN REVOLUTION

Dry sunny weather for ripening and harvest

Temperatures:
18° C for growth for 3 months
24° C for ripening

Lots of labour.

Flat land which will stop water draining away, allowing the rice to grow in it

Plenty of moisture for growth

Growing season:
only 90 days for quick maturing varieties
Most take 120 days or more

FIGURE 3.29
The growing requirements of rice

Increased output has created a surplus to sell in the cities. This has increased the farmer's quality of life, allowing reinvestment in the farm system with the purchase of machinery. These benefits have not been experienced by all farmers, however. Many poor farmers have not been able to afford the HYVs or the fertilizers, and therefore not had opportunities to improve their yields. Many farmers borrowed money to buy HYV seed and fertilisers but failure to pay it back has resulted in debt, forcing them off the land to the cities. As a result, rural inequalities have increased. As the wealthy farmers have become richer, the poor farmers have become poorer.

• Mechanisation

Tractors and mechanised ploughs were imported from MEDCs to use instead of water buffalo. Such technology was inappropriate on small farms, but most farmers could not afford to buy the equipment, let alone the fuel or spare parts, so only the wealthy farmers with large farms could mechanise to improve farm processes. This further reinforced rural inequalities.

• Irrigation

The monsoon rains are often unreliable. The new HYV seeds require far more water than the traditional varieties of rice. As a result, there has been an increased need to irrigate the land. About 45 million ha of land is irrigated in India. The traditional method of irrigation in the Ganges valley is digging wells. Holes are dug to reach the water table and each well can irrigate 1–2 ha of land. The Green Revolution is changing the traditional methods of lifting the water from the well by using modern electric or diesel pumps.

TASKS

1   Use an atlas to find the world's major rice growing areas. Plot them on an outline map of the world.

2   Explain why much of India is well suited to rice growing.

3   Why is rice cultivation in India so labour intensive?

4   Use the farming year diagram (Figure 3.28) to explain how rice growing is dependent on the monsoon climate.

5   Compare the photographs of subsistence rice growing (Figure 3.29) and the commercial dairy farm (Figure 3.25). Identify differences in the way the land is farmed.

India has also invested in large-scale water projects, such as the Narmada River Project which has built 30 dams in the river basin to irrigate 2 million ha of land and provide hydro-electricity to rural communities. The project has created much disruption, however, with more than 100 000 ha of forest and good agricultural land being flooded. Irrigation can lead to waterlogging of soils and salination, when salt moves upwards through the soil as moisture evaporates.

• Land reform

One of the fundamental problems holding back agricultural development in many LEDCs is the small size of many farms. In India, for example, 75% of farmers own less than 3 ha and 25% own less than 0.5 ha. This can be made worse by the fact that the farmer's land is often broken up into many tiny plots spread over a wide area. The majority of farmland is owned by a few wealthy landowners, who became even wealthier with the adoption of HYV varieties. Some of this wealth is spent buying more land, often from poorer farmers who run into debt from buying HYV seed and fertilisers.

India has attempted to introduce land reforms, aiming to increase average farm size, setting an upper limit on the amount of land held by the few wealthiest landowners, and has reallocated 2 million ha of surplus land to the landless. Such reforms have proved to be very unpopular with the powerful wealthy landowners, who are reluctant to lose their source of wealth. In many areas, little progress has been made, although in the state of Kerala, it has been successful.

• Appropriate technology

Trying to introduce the technology used in MEDCs as part of the Green Revolution has clearly created significant problems in many LEDCs. Now **appropriate technology** is seen as an alternative route to progress. Appropriate technology (sometimes called intermediate technology) aims to

help people in LEDCs solve their own problems. It is technology that is accessible and affordable to ordinary women and men to use in their own communities, and which is both economically and environmentally sustainable for them. In India, recent appropriate technology schemes have included training courses for animal first-aid workers, designing a fuel-efficient stove, designing and building improved fishing boats. Schemes focus on producing low-cost, easily manageable developments which improve people's quality of life. It avoids using hi-tech machinery that needs training, expensive fuel and spare parts imported from MEDCs.

TASKS

1 What is the Green Revolution?

2 Why was the Green Revolution needed?

3 Draw a table like the one shown below and list the advantages and disadvantages of the Green Revolution.

Advantages	Disadvantages

4 Why is land reform an unpopular aspect of the Green Revolution in India?

5 If you have access to the Internet, find out further information about:

  i International Rice Research Institute (IRRI)
    http://www.cgiar.org/irri/

  ii Intermediate Technology, a voluntary non-governmental organisation
    http://www.oneworld.org/itdg/

6 Read the FAO article on page 147 about the potential future crisis in rice production and then answer the following questions:

  a Explain why within 20 years most Asian countries will no longer be self-sufficient in rice.

  b Outline the reasons why Green Revolution technologies seem to be 'almost exhausted'.

  c What is meant by the term the 'greying' of Asia's rice farming? Explain why it is happening.

  d How can rice yields be increased in the future?

Rice crisis looms in Asia

Green Revolution technologies are 'almost exhausted' of any further productivity gains

By 2025, average rice yields must almost double, using less land, less water, less labour and fewer chemical inputs. Rice is the life-line of Asia. More than 90% of the world's rice total crop – currently some 520 million tonnes – is produced there, providing the region's 3100 million people more than a third of their total calories. With Asia's population growing by some 56 million a year, domestic demand for rice is expected to top 770 million tonnes by the year 2025.

How that increase will be achieved is the subject of growing concern among rice scientists and policymakers alike. If present trends continue, within 20 years most countries will no longer be self-sufficient in rice and Asia's legendary rice bowl will be filled increasingly by grain imports.

At the latest session of the International Rice Commission, held in Cairo this month, rice experts were describing the challenges ahead as 'mind boggling' and even 'frightening'. To meet rice demand over the next 30 years, the yield of irrigated rice in Asia will need to increase to about 6 tonnes/ha, nearly twice the current level. And this will have to be achieved using less land, less water, less labour and fewer chemical inputs, particularly pesticides.

Green Revolution technologies, which spurred increases in annual rice output of more than 3% – and probably saved millions from the threat of famine – are now considered 'almost exhausted' of any further productivity gains.

The size of Asia's rice lands is shrinking, under pressure from industrialisation and urbanisation. In China, the area under rice fell from 37 million ha in 1976 to 31 million ha in 1996. Further, soil salinisation, waterlogging and other degradation associated with intensive rice cropping may lead to a net drop in Asia's total irrigated area. Land suitable for further expansion of rice is also disappearing: water and wind erosion are estimated to affect some 400 million ha of the region's farm land, while another 47 million are subject to chemical and physical degradation. Over the next 25 years, uncropped land will be halved in South Asia and reduced by one-third in East Asia. The quantity and quality of water available for rice growing is also expected to decline.

Last, but not least, production will also to be affected by the 'greying' of Asia's rice farming population. The average age of farmers is increasing in almost every country, in parallel with its rate of industrialisation. In Korea, the number of rice farmers fell by two-thirds between 1965 and 1995. Urbanisation and industrialisation will further reduce the labour force, push up farm wages, increase farm size, and increase pressure for mechanisation.

Conclusion: even if the current level of productivity is sustained, it cannot match the food needs of Asia's expanding population. Some rice scientists say that only aggressive research aimed at breaking through present yield ceilings and establishing a new, stable yield plateau can 'help prevent a disaster'.

Hybrids and 'super rice'

The future research scenario will probably focus on hybrid rice and what is known as the New Plant Type (NPT), or 'super rice'. Hybrid rice is the only genetic yield-enhancing technology to have emerged since the Green Revolution. With yields up to 20% higher than those of conventional HYVs, hybrids have been widely adopted in China, where they now cover more than 50% of the total rice-planted area and account for about two-thirds of national production. However, transferring Chinese hybrid technology to other Asia countries has proven difficult, mainly due to the technical problems and costs involved in producing hybrid seeds. FAO and IRRI have now created a task force to promote development and use of hybrids in 12 other Asian countries. Meanwhile, work on 'super rice' is virtually complete at IRRI, and the plant promises to increase land productivity significantly. However, further intensive research will be needed to realize NPT's full potential of 15 tonnes/ha, and improve its disease and insect-pest resistance.

Biotechnology can also help, improving resistance to major pests and diseases, and transferring genes to rice from wild and unrelated species.

Finally, to be effective, strategies to increase Asian rice production must be supported by sound government policies and will depend heavily on adequate information on genetic resources, land use, water availability and irrigation potential.

Source: FAO: AG21: Magazine: Spotlight: Rice

http://www.fao.org/WAICENT/FAOINFO/AGRICULT/magazine/9809/spot1.htm

CHANGES IN FARMING IN AN MEDC

In the middle of the twentieth century, when Britain was faced with being starved into defeat during the Second World War, the government launched a campaign to increase agricultural production. After the war, politicians passed legislation that aimed to make Britain able to produce all the food it needed. From land drainage to hedgerow removal, there were generous government grants to improve farm efficiency.

In 1973, Britain joined the Common Market (now the European Union) and adopted the Common Agricultural Policy (CAP). The basic aims of CAP were to:

▶ Create a single market in which agricultural products could move freely.

▶ Make the community more self-sufficient by giving preference to EU produce and restrict imports from elsewhere

▶ Increase agricultural production through payments to farmers (often called subsidies). Under the CAP farmers are guaranteed fixed prices for their farm products within the EU, whatever the world market prices.

▶ Increase the average field size, farm size and farmers' income

The CAP costs nearly two-thirds of the European Union budget, even though farming only provides 5% of the EU's total income.

The CAP was so successful that by the mid-1980s there were large surpluses of most foods, the so-called cereal, butter and beef 'mountains' and the wine and milk 'lakes', which were costly to store and dispose of. The introduction of milk quotas helped to reduce overproduction of milk. However, the rising costs of the CAP and environmental concerns forced politicians into major CAP reforms in 1992.

The EU's response to the grain surplus has been **set aside**, introduced in 1992. This removed parts of farms from production for five years. The land can be left fallow, planted with trees or used for non-agricultural purposes, in return for money from the EU. Since 1995, support for cereals has been cut, in exchange for farmers being paid for their set aside land. In 1995, UK farmers were paid £253 per hectare for fallow, £140 per ha for cereals and £445 per ha for oilseed rape. The high subsidy for oil seed rape (Figure 3.30) was an incentive as the EU had a shortage of vegetable oil. Rape seed has a 40% oil content. It is used to produce cooking oil, margarine and salad dressing. Oil seed also improves the structure and fertility of the soil.

FIGURE 3.30
Oil seed rape

Productivity was also raised by technical progress in animal and crop breeding and increasing inputs from pesticides, fertilisers and machinery. **Agribusiness** is the increasing industrialisation of farming. Many farms have become bigger. New, larger machinery suits larger fields and needs fewer farm workers. Wetlands have been drained. Landscapes have changed dramatically and wildlife habitats and rural traditions have been lost.

1950

2000

FIGURE 3.31

A farm in 1950 and in 2000

TASKS

1 Why did farming in the UK start to change about 50 years ago?

2 What were the aims of the European Union's Common Agricultural Policy?

3 Why was the CAP too successful?

4 What is set aside?

5 How and why were EU farmers encouraged to grow oil seed rape?

6 What is meant by the term 'agribusiness'?

7 Look carefully at the field sketches of a farm in 1950 and 2000 (Figure 3.31).

   a Identify changes in the farming landscape which have taken place between 1950 and 2000.

   b Explain why the farming landscape has changed between these two years.

THE ENVIRONMENTAL IMPACT OF FARMING

Removal of hedgerows

Between 1945 and 1990, 380 000 km of hedgerows (nearly 50% of the UK's total) were removed. This was done by farmers to create bigger fields. Hedgerows take up space which can be used for growing crops. They also limit the size of machinery that can be used in fields. Hedgerows are also home for many harmful insects and pests, and the farmer has to spend valuable time and money maintaining them. Removal has had a major environmental impact, however. Soil erosion from wind and water has increased as the hedgerows are no longer there to act as wind-breaks. Wildlife has also been affected, as shown in the graphic opposite and news article below.

BBC News Online

Thursday, August 12, 1999 Published at 11:10 GMT 12:10 UK

Farmland birds in crisis

By Environment Correspondent Alex Kirby

A team of British researchers is calling for reform of Europe's agricultural policy to allow birds and other wild creatures a greater chance of survival. The team estimates that, in Britain alone, loss of biodiversity has meant the disappearance of 10 million breeding individuals of 10 farmland bird species in the past two decades. They say: 'Parallel changes have taken place in many other European countries.'

History repeats itself
In all, 116 species of farmland birds – one-fifth of European avifauna – are now of conservation concern. Intensification of agriculture 'is about making as great a proportion of primary production as possible available for human consumption. To the extent that this is achieved, the rest of nature is bound to suffer.'

They identify some key changes in British farming over the last 30 years:

- land drainage
- hedgerow removal
- introduction of new crop types
- increased use of agrochemicals, and a move towards monoculture
- a change from spring to autumn sowing

- the harvesting of grass for silage
- a reduction in the traditional rotation of crops.

The researchers say most of the evidence that farming is to blame for what is happening to the birds 'is by association, but in sum total it is damning'. 'Annual BTO censuses of 42 species of breeding birds show that 13 species living exclusively in farmland, such as the skylark and corn bunting, declined by an average of 30% between 1968 and 1995. 'Yet 29 species of habitat generalists, such as the carrion crow and the wren, have increased by an average of 23%.'

The researchers say there are agricultural schemes in the UK that could help biodiversity.

But there is 'no magic bullet with which to reverse the declines of a large suite of species … the most general prescription is to reverse the intensification of agriculture'.

Areas for research they recommend include the effects of introducing more variety into farming.

Chemical pollution

When we look at the countryside we don't always see the real problems. The countryside faces hidden perils from pollution. Changes in farming since the 1940s have led to increased water pollution. The agricultural industry is one of the four main sources of water pollution in the UK. Pollutants are washed from fields by rain water. Agriculture accounted for 13% of recorded pollution incidents in England and Wales in 1994.

Fertiliser run-off from farmland, contamination from silage and slurry (animal waste) all contribute to the over-enrichment of water by nutrients. Over-enrichment results in an overall decrease in the diversity of life in fresh waters. At the same time there is an increase in a few species which can withstand high nutrient levels, notably some algae. There is concern about pesticide use, especially in farming. This has led to the loss of wildlife caused by direct poisoning, risks to human health and increasing resistance to pesticides which, in turn, often means that more pesticides are used.

HEDGEROWS:
Life support for species at risk

CRISIS IN THE COUNTRYSIDE

One fifth of Britain's hedgerows have been removed in a generation, jeopardising birds, insects and mammals that have depended on them for shelter and flood since Saxon times.

Hedges trimmed into A shape support most diverse wildlife.

Flat-topped hedges offer less light.

Traditional cutting and laying of hawthorn and hazel declined as farmers wanted bigger fields.

Barn Owl

Hedgerow destruction depriving barn owls of prime food sources is significant factor in 70% fall in numbers since 1930s. Only 3,500 pairs remain.

Greater horseshoe bat

Bats follow hedgerows when flying between roosts and hunting grounds but become disorienteted if line is broken.

Wood cranesbill   **Ox-eye daisy**

Of nearly 300 plants recorded in hedgerows, many have vanished from surrounding countryside

FIGURE 3.32

Hedgerows at risk

TASKS

1 Why have farmers removed hedgerows?

2 What is the environmental impact of removing hedgerows?

3 Read the news article and explain why the researchers believe that 'farmland birds are in crisis'.

4 Look carefully at the cartoon (Figure 3.33). Explain what message it is trying to get across.

5 List the ways in which farming is damaging the environment. For each, identify measures that could be taken to reduce the damage.

FIGURE 3.33

Farmland birds are in crisis

THE LOCATION OF INDUSTRY

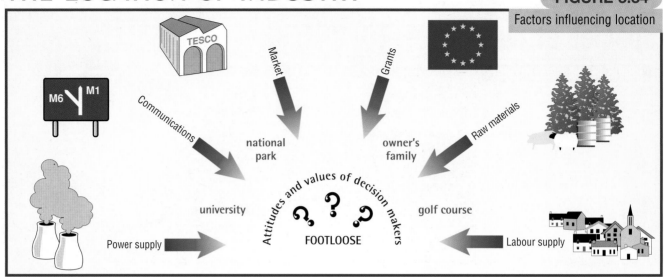

FIGURE 3.34

Factors influencing location

Choosing the location of a new factory is the result of a complex set of decisions. Usually, the aim is to locate where the most profit can be made. This can be influenced by all of the factors shown in Figure 3.34.

Sometimes one of these factors is particularly important. A frozen foods factory needs to locate close to its **raw materials** in a vegetable growing area so that the crops are fresh when they reach the factory. It would be an even better location if it were close to a fishing port. During the winter, the factory could concentrate on freezing fish. Although the frozen food may be sold all over the UK, it is neither difficult nor expensive to transport it. Vegetables and fish would be the factory's main raw materials and the most important reason for its location.

Alternatively, **labour** supply may be the dominant factor, such as in Birmingham's Jewellery Quarter, where special skills are needed. Making jewellery is one of the industries where computers have not taken over from human skill. In a few cases, the need for **energy** is particularly important. Page 154 looks at smelting aluminium. More often, being near the **market** for the product gives an advantage. It could be for freshness (supplying bread) or to keep distribution costs down (brewing beer).

Improvements in transport have had the greatest effect on where industry locates. At one time, heavy industry needed to be where coal could be unloaded from canal boats or railway wagons. Now most factories use electricity, which is available everywhere. Canalside factories still exist. They have stayed even though their main locating factor no longer applies. This is **industrial inertia**. The advantages of staying (local workforce, connections to suppliers and markets) are greater than the advantages of moving to any new location.

For a new industry, being close to a motorway junction is often very attractive. Many firms operate 'just in time' for their supplies. This means that rather than keep stocks of each component, they rely on them being delivered as they need them. This is how Ford operates its car factories. In such cases, **transport** becomes a more important location factor.

When a new factory is proposed, government help, or the amount of **grant** from the European Union may be the deciding factor.

Often it is impossible to identify which location factor is most significant as the interaction between them is complex. Most new industry today is **footloose**. Its success hardly depends on its location. The **attitudes** and **values** of decision makers mean choosing the location may even be based on where the owner or manager wants to live.

ARGOS DISTRIBUTION WAREHOUSE, STAFFORD

FIGURE 3.35

Argos, Acton Gate, Stafford

Argos opened its warehouse at Acton Gate, Stafford in 1998. They chose a site alongside Junction 13 of the M6, where the A449 trunk road crosses it. Easy access to the motorway network was the most important factor. But why this motorway junction in this part of England? It is a central location, so lorries can deliver overnight to the eleven regional bases throughout Britain and return within the number of hours drivers are allowed to work. A computer model was used by the decision makers.

Before Acton Gate opened, the company had a small warehouse in Penkridge, 5 km away, so they were able to retain their workforce. At busy times, such as before Christmas, many extra workers are needed. For labour supply, being close to the town of Stafford was a useful factor. Argos's 'raw materials' are the goods it sells. Their many origins were not a factor as the cost of transport to Stafford is paid by the manufacturers.

Acton Gate was a flat, greenfield site, so it was cheap to develop. It was not, however, a perfect location. It is too small and there is no room to expand.

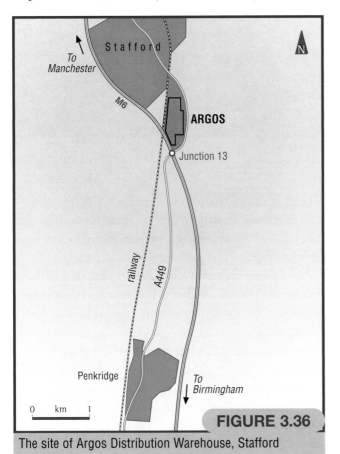

FIGURE 3.36

The site of Argos Distribution Warehouse, Stafford

TASKS

1 a Put in order of importance (most important first) the factors for choosing this location.
   b For each factor, explain why this location was chosen.

2 Look at Figures 3.35 and 3.36. Towards which direction was the camera pointing?

3 Look at the map (Figure 3.36). There is another new development to the north of the site.
   a What stops Argos expanding in other directions on this site?
   b A taller building was wanted but planners did not allow this. Suggest why.

ALCAN SMELTER, LYNEMOUTH

FIGURE 3.37

The Lynemouth smelter

Lynemouth is a village in Northumberland, chosen for a massive aluminium smelter. Aluminium is a light and flexible metal with many uses, from aircraft wings to the foil over the Christmas turkey. It is produced from bauxite, a rock found in some tropical regions. Australia, Guinea and Jamaica mine most of the world's bauxite. There is none however near Lynemouth. So why is the smelter here?

The choice of Lynemouth shows how several location factors interact. Usually the most important factor for an aluminium company is a supply of cheap power. To produce 1 tonne of aluminium uses as much electricity as a family uses in 20 years. The Lynemouth smelter has its own power station alongside. Next to that is a coal mine, supplying the power station directly.

There were many suitable sites for a smelter and power station in other parts of Britain and beyond. The reason for choosing Lynemouth introduces another factor. The government was concerned about unemployment and gave a grant of £28 million to create jobs here. For years, coal-mining was the main employer in this region. When the mines closed, there were very few alternative jobs for miners. Now the Lynemouth–Ellington Complex is the last deep mine in north-east England. Some of the coal comes from several kilometres away under the North Sea. Many of the smelter's workers are former coalminers. Like mining, aluminium production is heavy work and they are used to it.

As well as energy, labour supply and a government grant, other factors made Lynemouth a good location.

▸ There was a large area of flat land available.

▸ Aluminium is produced for firms in the UK, Ireland and much of Northern Europe. Lynemouth is central for supplying its market.

▸ Bauxite is bulky but the first stage of refining, making a white powder called alumina, greatly reduces the amount. So this is done near the bauxite mine and alumina is brought across the Atlantic Ocean to Blyth, a port originally built for exporting coal. It is only 13 km from Lynemouth and there is a direct rail link.

TASKS

1 For each factor that made Lynemouth a suitable location for an aluminium smelter, write in note form how it influenced in the decision to build here. You can present it as a star diagram.

2 Look at Figure 3.37. Use Figure 3.38 to work out which direction the camera was pointing towards.

3 Draw a sketch of Figure 3.37. Label:
 a the aluminium smelter
 b the power station
 c the coal mine
 d the railway
 e the North Sea.

4 a Why is bauxite not brought to Lynemouth?
 b Suggest why not much aluminium is produced where most bauxite is mined.

5 If the coal mine closed, the smelter would probably continue to produce aluminium here.
 a Why is this likely?
 b Explain how this would be possible.
 c What term would describe the smelter's location if the mine closed? (Clue: look at page 152.)

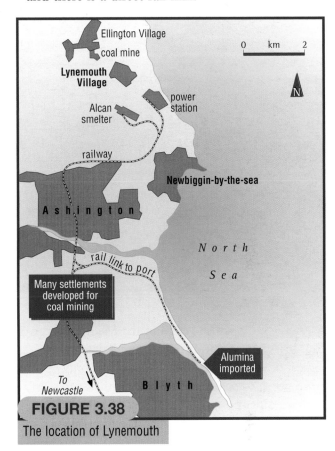

FIGURE 3.38

The location of Lynemouth

FIGURE 3.39

Ellington Colliery

THE EFFECTS OF TOURISM – MENORCA

FIGURE 3.40
Menorca holiday photo

Menorca is a Spanish island in the Mediterranean Sea. It is part of a group called the Balaerics, the largest of which is Majorca. Menorca is only 19 km from north to south and 48 km from east to west, with a resident population of just 69 000. It is a holiday island and, in summer, the population more than doubles. Three-quarters of visitors are British; most of the rest are Germans and Scandinavians.

▶ As incomes have risen in these countries, more people can afford holidays – and can afford to travel further from home.

▶ Employers now give their workers more time off.

▶ Charter flights have made flying cheaper.

▶ A new airport with a longer runway was opened.

▶ Jet aircraft have made journeys faster. Menorca is about two hours' flying time from the UK.

▶ Package holidays have made it easier to go abroad, as travel and accommodation are arranged for the tourist.

Menorca has been a tourist destination for less than 50 years. Hotels and apartments grew rapidly and now tourism is the largest employer.

FIGURE 3.41
Development in Menorca

In 1960, agriculture generated 50% of Menorca's wealth; now it is only 1.6%. Farming, though, has done well from tourism, supplying the hotels. Craft industries too have benefited from selling to tourists, who take home Menorcan jewellery, leather goods and gin. So there has been a **multiplier effect** to the development of tourism. It brings extra business to taxi drivers – and making ice cream is a major Menorcan industry.

Although Menorca has hot, dry summers, winters are warm and wet. Between November and April there are very few tourists, so there is a shortage of all-year employment. The government provides *paro*, money for tourism workers who are unemployed during winter. Every autumn, some workers leave Menorca until next year's tourist season starts. Even so, at 9% for males, unemployment is the lowest of any Spanish region and the average income is 60 per cent above the

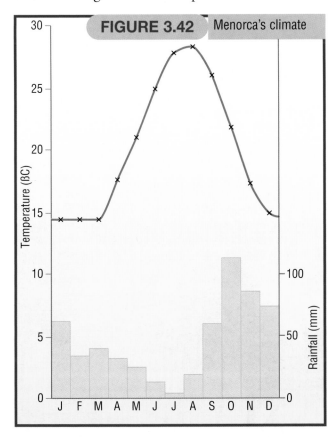

FIGURE 3.42 Menorca's climate

country's average. Before tourism developed, many young people, especially men, left the island permanently to look for work in mainland Spain. Now the population is rising by about a thousand people a year. Some of these are foreigners, coming to retire. They may change the way of life of the

Menorcan communities they join, or keep themselves separate. There is some local resentment over villas bought by foreigners.

The government is concerned about preserving the language. All place names are now in Menorqui, the island's traditional language, and the use of foreign words to advertise to tourists is controlled. Fines can be as high as 2.5 million pesetas (£10 000).

After the rapid growth of tourism, Menorca is learning to live with slower and more careful development. There is more awareness of the need to protect the remaining natural resources. Some of the early hotels are eyesores, spoiling beautiful coastal scenery. Now, within 250 m of the coast, new buildings cannot be higher than two storeys. Undeveloped areas are to remain so. The United Nations has declared the whole island a Biosphere Reserve, to acknowledge the importance of conserving Menorca's natural environment. A bridle path round the entire coast is to be promoted to encourage 'quiet recreation'. Many Menorcans believe that development has gone as far as it should go.

TASKS

1 Suggest some reasons why Menorca is a popular choice for tourists from UK, Germany and Scandinavia. Think about wealth, transport and climate.

2 Describe the attractions for tourists shown in Figures 3.40 and 3.41.

3 List some of the jobs that have been created by the growth of tourism.

4 Use an example from Menorca to explain what the multiplier effect is.

5 Explain why Menorca attracts few tourists between November to April. Support your answer with information from Figure 3.42.

6 What problems has tourism brought to Menorca?

7 What are the advantages and disadvantages of encouraging the use of Menorca's traditional language?

8 'The environment is Menorca's most valuable raw material and it is non-renewable' said a politician arguing against permission for new hotels. What is Menorca doing to protect the environment?

You can find out more about the attractions of Menorca from the website of a UK holiday accommodation agency on the island. You can use this information to help you complete the tasks.

http://www.menorca.co.uk/

THE EFFECTS OF TOURISM – KENYA

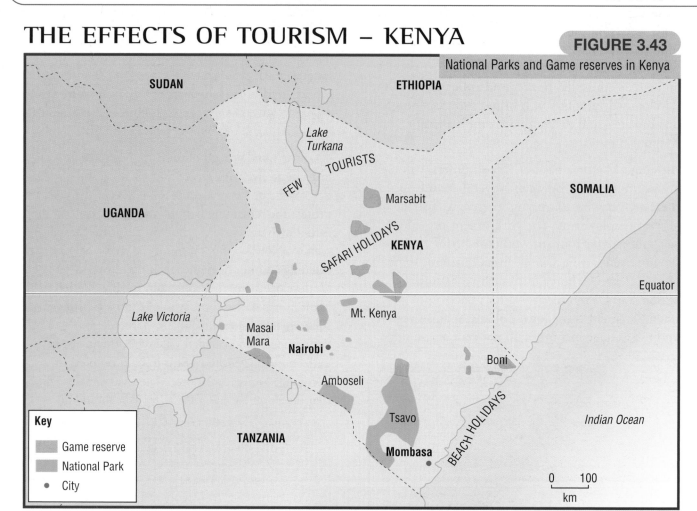

FIGURE 3.43

National Parks and Game reserves in Kenya

Mass tourism came to Kenya 30 years later than Menorca. Larger aircraft with longer range fuel tanks made charter flights possible – and at prices not much higher than a package holiday in Menorca. Many people who had enjoyed Mediterranean holidays were looking for something different and holiday programmes on television were starting to feature destinations in LEDCs such as Kenya.

Kenya offers an attractive climate, with hot sunshine all year round. There are beach resorts on the coast of the Indian Ocean with fine sand and coral reefs to explore (Figure 3.44). Inland, there are vast areas of savanna grassland set aside as national parks and game reserves (Figure 3.45). As well as seeing wild animals, another attraction is to experience a different culture. It is common to follow a week's safari with a week at the coast, staying in European-style hotels.

FIGURE 3.44

A Kenyan beach holiday

Benefits

As well as building hotels, money has been spent on developing Nairobi and Mombasa airports and improving roads to the game reserves. Some of the improvements for the tourists have benefited other people too. Mombasa now has a reliable water supply, for example. Tourism has created jobs for Kenyans and, unlike in Menorca, they are all the year round. In less than ten years, tourism became Kenya's biggest earner.

Problems

The benefits have not been spread evenly.

▶ Most new jobs are in the better off cities, which encourages more migration and adds to the shanty towns.

▶ Nearly all the jobs are unskilled and poorly paid.

▶ Most of the money available to the government has been spent in the cities too.

▶ More than half of the money paid for the holidays never comes to Kenya. The travel companies, airlines and many hotels are foreign owned.

▶ In places, the environment has suffered badly from sewage in the sea and safari minibuses damaging the savanna ecosystem.

▶ People who used to live and keep their herds in Kenya's national parks have been forced out, losing their land and their traditional way of life.

▶ Seeing the wealth of tourists has led to resentment among the poor and threatens the traditional culture. Drugs and crime have increased and AIDS is a major problem.

▶ Tourist destinations go in and out of fashion. During the 1990s, visitors to Kenya from the UK decreased, as a result of fears over tourists' safety.

TASKS

1 What attractions does Kenya have for tourists?

2 What developments led to the growth of tourism in Kenya?

3 Explain how the development of tourism has benefited life in Kenya.

4 Look at Figure 3.45. Describe the vehicles used for safari holidays.

Kenyaweb, a website based in Kenya, provides further information about the tourist attractions of the country. You can obtain more data to help you complete the tasks.

http://www.kenyaweb.com/

FIGURE 3.45

Safari traffic jam when a lion is sighted

CHANGING ENERGY USE IN THE UK

In most MEDCs, the demand for energy continues to increase but the source of this energy will change as fossil fuel reserves decline and alternative types of energy are developed (Figure 3.46).

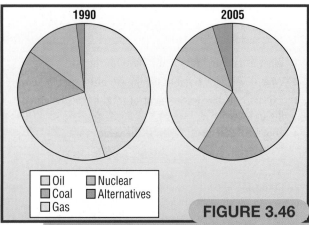

FIGURE 3.46

Types of energy use in the UK in 1990 and 2005

The coal industry

The Industrial Revolution in Europe was based on coal. European coalfields became the centres of

FIGURE 3.47

UK coalfields

industrial and urban development. The UK has several major coalfields (Figure 3.47). Estimated coal reserves could have kept Britain supplied for another 200 years.

The UK coal industry has been in decline since the 1920s when competition resulted in the loss of world markets. During the 1930s the lack of investment and strikes threatened the industry. In 1948 the coal industry was nationalised and the government modernised the pits and supported mining communities. In this process, numerous smaller, unprofitable pits closed, replaced by larger 'super' pits. Despite considerable investment the coal industry failed to make a profit. The situation was not helped by industrial disputes, for example that in 1984.

In the early 1990s, the coal industry suffered a rapid decline as the government decided to close many pits in readiness for privatisation of the industry. The sell-off took place in 1994. Since then, mines with limited coal reserves and geological problems such as thin seams, have also closed. There has been a considerable social cost and economic decline as coalfield communities lost employment.

Year	Amount of coal mined (m tonnes)	No. of mines	No. of miners	Average amount coal produced per miner (m tonnes)
1950	220	901	688 000	320
1960	197	698	588 000	335
1970	145	292	286 000	507
1980	126	211	230 000	548
1990	81	73	65 000	1462
2000	43	13	7 000	1600

Figure 3.48 The British coal mining industry

The coal industry declined because it became harder to sell coal. Most was sold to power stations, but the requirement that UK power stations should buy British coal was withdrawn in favour of buying from the cheapest source. The electricity generating companies decided to import cheaper coal and build new gas-fired power stations that used natural gas. The coal industry had lost its main market (Figure 3.49).

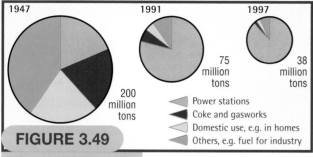

FIGURE 3.49

Uses of coal 1947–1997

1947 — 200 million tons
1991 — 75 million tons
1997 — 38 million tons

- Power stations
- Coke and gasworks
- Domestic use, e.g. in homes
- Others, e.g. fuel for industry

Environmental concerns over the damage caused by mining and clean air legislation also affected coal's image as a fuel. International agreements about pollution and the threat of acid rain led to strict limits on CO_2 and SO_4 were agreed. Some coal burning power stations installed expensive filters to reduce emissions. Alternative cleaner energy sources such as wind power are being developed.

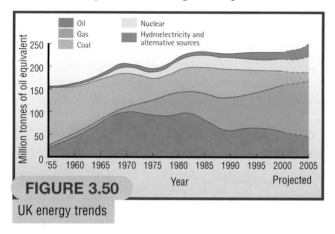

FIGURE 3.50

UK energy trends

- Oil
- Gas
- Coal
- Nuclear
- Hydroelectricity and alternative sources

Oil and gas

The UK has extensive oil fields in the North Sea and Irish Sea. This helped the UK reduce its oil imports. Together with natural gas, oil has replaced coal as Britain's main energy sources. Oil was cheap and was in demand from industry and transport (Figure 3.50). Some coal continues to be mined from the remaining pits. But one-third of coal will be recovered from open cast sites, for example in north-east Derbyshire (Figure 3.51).

FIGURE 3.51

Open-cast site in Derbyshire

FIGURE 3.52

A modern gas-fired power station at Sutton Bridge, Lincolnshire

Small amounts of natural gas are burnt off as a waste product from oil fields. When large reserves of gas were discovered in the North Sea in 1965, it became worthwhile to use this valuable energy resource. The consumption of natural gas has increased steadily to 32% of UK energy needs. In the 1990s, gas rather than coal was chosen to generate electricity in new power stations. This became known as the 'dash for gas'. Natural gas is a cleaner fuel than coal and its use helped the government meet international pollution agreements. The large reserves meant it was, to begin with, a cheaper fuel than coal. Some coal-fired power stations were converted to gas, but several new stations have been built, for example, Sutton Bridge, Lincolnshire, (Figure 3.52).

TASKS

1 Describe how production of energy has changed in the UK.

2 Plot the data in Figure 3.48 to show the decline of the coal industry.

3 Make a list of the reasons for the decline of the coal industry. Put * against the most important reason.

4 In recent years, open-cast mining has become more important. Describe how open-cast mining affects an area.

5 What factors have encouraged the 'dash for gas'?

6 Natural gas is a finite resource. Do you think the 'dash for gas' is a good idea? Explain what you see as the advantages and disadvantages of this policy.

MEETING FUTURE ENERGY DEMANDS

At present, less than 1% of the UK's energy is produced by alternative sources such as wind or HEP. Such sources are often expensive to establish but are cheaper to operate. They are based on renewable sources (Figure 3.53). The government has committed the country to achieving 4% of its needs from alternative sources.

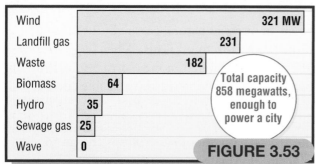

Total capacity 858 megawatts, enough to power a city

FIGURE 3.53

Electricity produced in the UK from renewable resources

Nuclear power

Nuclear energy provides 28% of UK's needs, but there are concerns over its safety. Britain first developed nuclear power during the 1950s and, at its peak, there were 38 nuclear power stations in service.

Nuclear power stations tend to be located on flat coastal sites. These are remote from centres of population (Figure 3.54). The process uses vast amounts of sea water for cooling. The fuel used is preprocessed uranium. Heat is produced from the nuclear fission. The heat turns water into steam. This steam is used to drive turbines that generate the electricity.

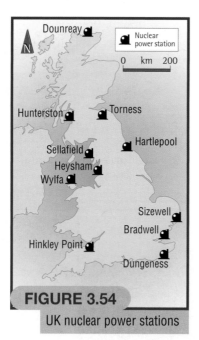

FIGURE 3.54

UK nuclear power stations

There is, however, opposition to the use and expansion of nuclear power. Several pressure groups raise concerns about its safety. They point to the Chernobyl disaster in 1986 when a reactor in the Ukraine exploded killing 31 people; 135 000 people were evacuated as the area around the plant became contaminated with radiation and will remain so for a very long time. Wind-blown radiation from the explosion came to Britain. Polluted rain made soil radioactive in parts of Wales and Scotland. Some farmers in Wales still have a ban on the sale of their sheep.

As well as the safety of the power stations, there is concern about nuclear waste. The Irish Sea has been polluted from waste from Sellafield in Cumbria. Nuclear fuel waste must be stored in a safe environment, possibly in underground facilities built into impermeable rock. It stays radioactive for thousands of years and needs to be stored very carefully.

Wind power

A safer but still controversial energy source is wind power. Britain has nearly half of Europe's wind potential (Figure 3.55).

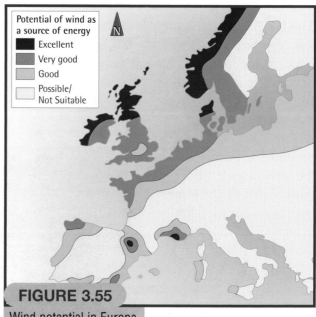

FIGURE 3.55

Wind potential in Europe

Electricity is generated by large blades turning a turbine which run a generator. To be worthwhile several wind turbines need to be located together on a **wind farm**. By 1996, there were 550 wind turbines in Britain. Large wind turbine farms are being developed in Spain, Portugal and Italy benefiting over 500 000 people.

FIGURE 3.57

A wind farm in mid-Wales

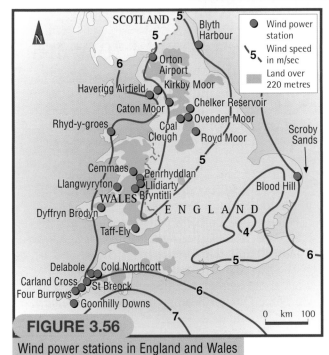

FIGURE 3.56

Wind power stations in England and Wales

Protestors believe that wind power sites disrupt remote places and provide very little electricity. Apart from their physical appearance, they produce a low thudding sound when operating which can be heard up to 10 km away. Opposition to wind farms often comes from those who support alternative energy sources in principle but do not want the sight and hum of aerogenerators near where they live (Figure 3.57). There is plan to build a wind farm with 25 turbines off the Norfolk coast on Scroby Sandbank, near Great Yarmouth, seeking to use offshore winds. Despite being away from people, there are concerns for sea birds and basking seals.

A wind farm usually consists of at least 20 turbines set on concrete bases. The turbines are expected to last 15 years. They only work when the wind blows and computers are used to gain optimum performance. The energy they produce creates no air pollution and uses very little land. A wind farm typically covers less than 2 sq. km. Most wind farms are located on highland in the west of Britain (Figure 3.56). They need an average wind speed of 5 km per second to work. However, during high winds the turbines have to be shut down.

TASKS

1 Describe the location of UK nuclear power stations. What do most of these sites have in common?

2 What are the advantages of nuclear power over fossil fuels?

3 Why are people concerned about nuclear power plant?

4 Study Figure 3.55. Identify the parts of Britain which have the best potential for wind farms.

5 Describe the location of the UK wind farms. How many are located in beautiful areas, like national parks?

6 Explain why four out of five people support wind power, yet there is considerable opposition to wind farms.

HOLMEWOOD

Holmewood is on the North Derbyshire coalfield (Figure 3.58). Small amounts of coal had been mined here since the Middle Ages. Most houses in Holmewood were built in 1901–08 for miners who worked in the pit, when new shafts were sunk. The population of 776 in 1901 had trebled by 1911. The pit produced coal which was distributed by railways and supplied a local brickworks.

The pit closed in 1970 and the area lost 2299 mining jobs. This was the pattern of decline across the North Derbyshire coalfield.

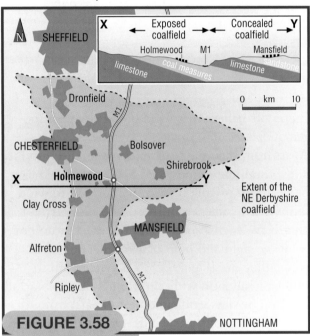

FIGURE 3.58
Location of Holmewood

In 1985, 34% of people who lived in north-east Derbyshire area still depended on the mining industry for employment. Since then, all the deep mines have closed and many more jobs have been lost in other traditional industries, such as iron and steel, chemicals, engineering and railways. It has left serious economic, environmental and social problems (Figure 3.60).

There were few alternative jobs to turn to. A few miners were **redeployed** to collieries outside the region. Apart from direct job losses in mining there were also indirect job losses in the supporting industries. It is estimated that for every 100 mining jobs lost a further 50 other jobs were also lost. For

example, in north-east Derbyshire, British Coal was supplied by over 60 firms in the Chesterfield area.

Economic costs

In 1987, Holmewood had an unemployment rate of 18%. That compared with a north-east Derbyshire average of 14% and a national average of 10%. Long-term male and youth unemployment has remained higher than in other regions since the pits closed.

Unemployment means that families have less income to spend which also means there is less demand for goods and services from local businesses. Holmewood like many other colliery villages lost shops and services as trade declined.

Holmewood pit closure is to be brought forward

Geological faults in the main coal seam have brought forward the closure date.
The life of the pit, which employs 800 men, was very limited, the spokesman said, because of the near exhaustion of workable coal reserves.

A closure date is to be announced within the next few weeks, but the men would be offered jobs at other North Derbyshire collieries, he said.

Holmewood Colliery, which produces about 6,500 tons a week, was said three years ago to be a short-life pit, due for closure by 1970.

The last coal face in the main seam will be worked out by the end of September, leaving only the Three-Quarter Seam, which is uneconomic and difficult to mine.

Figure 3.59 Newspaper extract (based on the *Sheffield Telegraph* report, 1968)

Social and environmental costs of closure

The social cost of pit closure was matched by the legacy of environmental problems. Poor housing that lacked basic amenities, blighted and derelict landscapes and a substandard **infrastructure** left the area unattractive to new industry and commercial development. Mining had left spoil heaps of waste rock, subsidence where the land had sunk and pollution problems.

Local councils only had limited resources to make an impact on these areas. Many sites lay derelict years after closure, waiting for government grants for reclamation schemes. For example, Langwith Colliery closed in 1978 but reclamation work did not begin until 1986.

The social costs of closure are hard to quantify. Indicators such as percentages claiming benefits, children on free school meals and those living in rented accommodation are higher than the national average. Also, there is less car ownership and fewer local shops. Some families, unable to find work locally move away leaving housing vacant. This adds to the sense of decline. Research has linked unemployment with poor physical and mental health. There are high levels of illness and increased death rate due to stress and poor diet. Unemployment can also cause social isolation and low self-esteem.

Redevelopment

Apart from reclamation, local authorities used a range of grants and initiatives to revitalise the former mining communities. Industrial estates located in **enterprise zones** encouraged new industry to set up. The Holmewood Enterprise Zone (Figure 3.61), advertises its improved infrastructure and links with the M1. The zone offers a range of tax and financial incentives such as no rates to pay for ten years, and fast-track administration and planning procedures. The landscaped area had purpose-built units. The aim was to create 60 new jobs per hectare.

FIGURE 3.60
Effects of pit closure

The industrial estate is growing and becoming established. However, Holmewood itself provides only 23% of the workforce and unemployment here remains above the national average. Recent private housing has been bought by commuters who work in Sheffield, Chesterfield and Nottingham.

TASKS

1 Describe how north-east Derbyshire industry was seriously affected by pit closures.
2 Describe why the rate of unemployment remained high following the pit closures.
3 Why do pit closures have a domino effect on an area?
4 List some possible environmental effects of mining.
5 Why are the social costs of pit closure hard to measure?
6 How can industrial estates and enterprise zones help a former pit village recover?

FIGURE 3.61
New industry in Holmewood Enterprise Zone

CONSETT AFTER COAL AND STEEL

FIGURE 3.62

Consett's improved transport infrastruture

Consett is a town of 25 000 people, high on the edge of the Pennines in north-east Durham. In 1841, it was a village community of only 145, but about to become a boom town. Below ground was coal and ironstone. Nearby was limestone. These were the three ingredients needed for blast furnaces to produce iron and steel. Thousands of migrants came to find work. The iron company looked after its workers, providing houses, a hospital, schools and a library. It did not have to pay high prices for coal as it owned 37 coal mines, so it could make iron at a good price. In less than 30 years, it became the largest ironworks in Britain.

As early as 1852, local ironstone was running out. It was cheaper to bring it from the Cleveland Hills in Yorkshire. Later it came from Kiruna in Sweden. This was the first sign of **industrial inertia**. (page 152).

A report in 1970 identified the best locations for making iron and steel as large, flat sites at the coast next to deep water ports. Consett's future was to change as rapidly as it had done 140 years earlier. In 1980, the steelworks closed and 3 715 people lost their jobs. As well as the steelworks, thousands more jobs were lost in other businesses such as transport and local shops. This is the **multiplier effect**.

By 1981, Consett was the worst employment blackspot in England, with 26.6% of the working population looking for jobs. Many had been steelworkers since leaving school and had few other skills. Many of the young moved to other areas to seek work, just as their ancestors had come to Consett in the nineteenth century, but young people gave life to the town.

The need was to diversify, to bring in new industry. The Government gave money to attract firms. The road network, the local environment and services were improved. By 1992, the rate of unemployment was down to 11%, but still more than twice the national average. In April 1992, 'Project Genesis' was announced.

Project Genesis

The first stage was to reclaim the massive site, as big as 280 football pitches. Not all is for new industry; some is being used for recreation and some has gone back to agriculture. There is public open space and land for horseriding. Large areas are set aside for new factories in parkland setting. Firms that are now established in Consett make crisps, caravans and Christmas decorations!

Project Genesis proposed a hotel to boost tourism, new shops to enlarge the town centre, and an extension to the bypass to make pedestrianisation possible. To the south of town, land was identified for a wind farm and a power station burning forest waste.

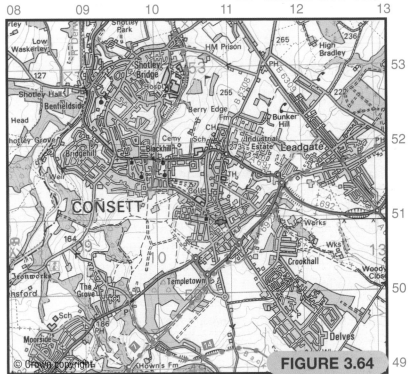

FIGURE 3.64

OS Landranger map of Consett

FIGURE 3.63

Pleasant surroundings for making Phileas Fogg crisps

TASKS

1 Find Consett in the atlas.
   a What direction and distance is Consett from Newcastle?
   b Use the atlas to describe the situation of Consett.

2 a List the reasons why Consett became an important iron and steel town.
   b For how many years were *all* its raw materials available locally?

3 a Describe what happened to Consett in the 1980s.
   b How did the multiplier effect work in Consett?

4 a Look at the Ordnance Survey map (Figure 3.63). Part of the steelworks was in the Templetown area of Consett. What is the land now used for?
   b Figure 3.62 was taken from grid reference 111506 looking towards the northeast. What is the number of this road?

c This new road has been built along the route of an old railway which used to bring coal and iron ore to Consett. Most of the old railways here are now shown with a green symbol. Use the key to an OS Landranger map to find out what this means.
   d What map evidence is there that tourism is being developed in this area?

5 What are the advantages of having a wide variety of industries in a town, rather than most employment being at a single large employer?

6 How has the quality of life been improved in Consett?

7 Who should pay for schemes like Project Genesis?

COURSEWORK FOR ECONOMIC ACTIVITIES
Farming

Opportunities for fieldwork are relatively limited. They are mainly restricted to those who live on farms, as farming contacts are essential. Coursework based upon just one farm rarely reaches the standard needed for grades A and B, mainly because of the shortage of primary data collection. The best chance of good coursework from a single farm study is for one made up of very different physical regions, such as a hill sheep farm with land on the valley floor, slopes and hill tops (Figure 3.65). Undertaking physical fieldwork, such as recording the weather, taking soil samples and studying the vegetation, is likely to produce varied results through which differences in land use can be explained. Modern farming methods have reduced the effects of physical variations on land use on most lowland farms. If two or more farms are compared, the chances of collecting sufficient data are increased because physical features and policies of the farmers can be expected to be different. Any study set up to explore changes over time matches GCSE content really well. Provided your farmers are co-operative, you can study farm records from previous years. Your best chance of making a success of the study is after a significant change on one of the farms, such as diversification.

FIGURE 3.65

Lake District farm. If only sheep could be interviewed to increase primary data collection from farm studies!

Manufacturing industry

A Inner city canalside location

B City edge location near a motorway

FIGURE 3.66

Two industrial locations suitable for a comparative study

Companies are notoriously uncooperative about divulging data. If you write letters to firms, some will answers questions, such as what they make, how many they employ and why they chose to locate there, but some don't reply. Others send glossy brochures or give vague replies, both of which border on the useless. One really useful piece of information is lists of post codes (and nothing else) from the home addresses of the workers, from which the factory's sphere of influence for employment can be mapped, but it is almost impossible to obtain.

Without a company contact, the best advice is to set up the study which is based upon observation such as:

▶ an investigation into the distribution of industry in town P;

▶ in what ways and why are the locations of old and new industrial areas different in town Q?

▶ what are the similarities and differences between industrial areas R and S?

Observation can be supported by an environmental quality survey and use of maps. Wider references to factors for industrial location and urban zones can also be made to incorporate the broader geographical background.

Tourism

Tourism and leisure are popular topics for coursework. Many are based in coastal resorts and use one or more of the following enquiry ideas.

which have appeared in most towns (Figure 3.67)? This means there are likely to be opportunities in a town near you. In towns and cities similar coursework can be done by using parks, sports grounds, country parks and leisure centres.

Much coursework in leisure and tourism depends upon collecting data by using questionnaires.

Tourists are often willing to spend more time answering your questions than locals. It is vital that your questionnaire is as good as it can be.

Enquiry idea	Fieldwork possibilities	Secondary sources
The distribution of tourist attractions and facilities and the reasons for the distribution.	Observation of the physical attractions e.g. beach, cliffs and the human facilities.	With the help of a large scale map or plan, map their locations.
How popular are the different attractions? Why are some more popular than others?	Do people counts. Give out questionnaires asking people where they have visited and why. Do an environmental quality survey. Examine ease of access e.g. car parks and public transport	Visitor figures from the Council or Tourist Information Office. Local bus routes and timetables.
Comparison of spheres of influence between two beaches or two or more attractions.	Observation of their physical / human attractions. Questionnaires to visitors handed out at the attractions.	Leaflets about them and publicity. Use maps and timetables to study accessibility.

FIGURE 3.67

Signpost outside York Minster to some of the other tourist attractions

Advice

Don't dream of doing a leisure centre study if all you want do is describe its facilities. This is not a geographical study. You need to concentrate on studying its sphere of influence.

You don't need to live in a coastal resort. Many inland towns attract significant numbers of day visitors. Have you noticed the visitor signposts

Do a draft version and pilot it by giving it to two or three people before making copies of it. Figure 3.68 shows a visitor questionnaire used in York, which worked quite well. This forms a good basis for other visitor questionnaires, but remember that the questionnaire perfect for your own study will never be printed in any book!

VISITOR QUESTIONNAIRE

1. In which town (in the UK) or country (foreign visitor) do you live?
2. How have you travelled here (from within the UK)?
   Car   Bus   Train   Cycle   Other
3. How long are you staying?
   Less than 1 day   2–3 days   5–7 days   1–2 weeks   Longer
4. In what type of accomodation are you staying?
   Hotel   Guest house/BB   Friends/Relatives   Camping/caravan   Other
5. Rate the following on a scale of 1–5   (1= poor, 3 = average, 5 = good)

	1	2	3	4	5
York Minster					
Jorvik Museum					
The Walls					
The Shambles					
The Castle Museum					
Shopping					
Eating					
Parking					
Entertainment					
Service by the local people					
Tidiness					

FIGURE 3.68

Visitor questionnaire used in York

EXAMINATION TECHNIQUE – UNIT 3

Unit 3 is about how people live in different areas of the world and what they need to carry on their lives. You will have studied topics on farming, industry, tourism and energy in LEDCs and MEDCs. The emphasis in examination advice from this unit will be on how to improve case study answers in Papers 1 and 2 . For many candidates, the case studies from this unit are the most difficult in which to score high marks. The questions from Papers 3 and 4 will concentrate upon dealing with graphs.

The **case study** is usually the final section of a question and often has the largest mark allocation: about 4 marks on Paper 1 and 7 marks on Paper 2. Throughout this book you will be learning about case studies. A case study is just a real world example of the **concept** you are studying. See pages 233 and 238 for some examples of the different ways in which case study questions can be set out in the examination paper.

In asking you to describe, give details about, or refer to an example that you have studied, the question is demanding that you know some details about a particular topic in a specific part of the world. The examiner will be looking for facts about the topic, not vague statements that might describe any industry or farming system in any part of the world. The 'secret' to scoring high marks on the case study question is to learn your notes in order to remember the details about why people grow rice in Bangladesh, or how tourism may be threatening the environment of Dartmoor.

This first example of a case study question focuses on farming in LEDCs. Although slightly different in style, the question is testing the same case study knowledge in both tiers of paper.

Foundation Tier question

1 (a) Farming has changed in many LEDCs. Choose an example to show how farming has changed in a named LEDC.

[4]

Higher Tier question

2 (a) For a named LEDC, describe how farming is changing. (7)

Notice how the mark allocation is different between the Foundation and Higher Tier questions. Both questions ask you to name your chosen example, in this case, a country. Be careful that you do not get mixed up between an LEDC and an MEDC – on many topics you will have an example of both. The questions are then straightforward but there are two points of which to be aware. The **command word** is 'describe' (see page 232 for details about command words). The main focus of the question is to describe farming *change*, not just what farming is like. Your answer must, therefore, describe the new developments which have occurred in your chosen country.

The following answers show how high marks can be scored on this question.

Foundation Tier answer

Name of LEDC: Brazil
Use of more machinery such as tractors to harvest more crops.
More healthier crops due to better irrigation.
More fertilisers are used to help the soil regain nutrients faster.

Examiner's comment

The farming changes which are described could apply to other LEDCs but they are all important developments in farming. For each change the answer is developed by describing its results.

Higher Tier answer

An LEDC where their farming methods are changing is India. In India there is the Green Revolution where developed countries are trying to help India develop modern techniques of farming such as the seeds which increase yield and double harvest. The developed countries wanted to help by developing mechanisation which would be more efficient and quicker than manual labour. The new seeds need plenty of fertiliser and pesticides and extra irrigation so Indian farmers have to provide them. It has cost them a considerable amount of money which they cannot afford.

Examiner's comment

Although rather a dated example, the Green Revolution represents an important series of changes that have occurred in many LEDCs. The changes are well developed by reference to the results, the problems and the dependence on help from MEDCs.

Unlike questions that are testing your geographical skills and understanding, case study questions do not usually refer to any kind of **resource** in the examination paper. The following question taken from the Foundation Tier paper is an exception in that there is some information to read before you begin to use your case study knowledge.

3 (a) Study Fig. 1

> Town devastated as factory is closed

> **Large company plans to open new store in town**

> **New quarry to be developed at local site**

> **End of an industrial era as last mine is closed down**

Describe how developments like these will affect the people of an area. Refer to an example which you have studied. **[4]**

The command word is again 'describe', but the instruction does not say that you have to write about all the examples which are given, rather that you must think about such types of development. You need only refer to one example which you have hopefully learned as your case study.

Look how the two answers below are different. One candidate makes a good attempt by focusing on the case study, whilst the other candidate jumps around from one idea to another.

Answer from candidate 1

Example studied: Cotgrave, Nottinghamshire

Cotgrave was once a small mining town but the coal mines have closed down because there is no more coal left to dig up. After they closed the mines many people were left unemployed. Many of the unemployed had to move near to the city to find jobs. Near the coal mine which has been closed down they have built a new industrial area.

Examiner's comment

The candidate chose a local example of industrial change and described two effects on local people. The answer could have been developed further by describing how the new industrial area might have affected local people.

Answer from candidate 2

Example studied: York

Development such as these cause a lot of pollution. When industries like mines are abandoned they make the environment ugly. Also when factories are closed it sometimes causes financial loss and in some cases unemployment.

Examiner's comment

The answer is not about York and is too general. Pollution, for example, is not explained. The other ideas are acceptable but too vague, they could apply to any industrial town.

The following question is taken from the Higher Tier paper and illustrates a very common question layout.

4 (a) For a named and located example of an industry within the UK which you have studied, explain why the industry has developed there. **[7]**

First of all the instruction is to name and locate your case study. Then you must explain why it grew up there (see page 232 for further guidance on **command** words). A mark allocation of seven marks clearly expects a detailed answer. Think about why the following two candidates had different degrees of success in answering this question.

Answer from candidate 1

In South Wales there was a large iron and steel industry. It grew up there because:

– there was a readily available workforce.

– that workforce was used to the Welsh climate.

– there were large deposits or iron ore there which was used for making iron and steel.

– there were also large amounts of limestone, also needed for iron and steel production.

– it has good transport connections via railways and roads.

– it had large supplies of coal which could be mined easily.

– it was close to a major city (Cardiff) where they could sell their goods.

– the coast was close so they could sell abroad.

Examiner's comment

The candidate gives a knowledgeable answer about a specific industry in a named area. The different ideas are simply explained as a series of bullet points. Some of the ideas needed further explanation, for example the road and rail connections.

Answer from candidate 2

One industry is Parkhouse industrial estate. This is an ideal site for industry. The land is flat and spacious. There is room for further expansion if needed. The industrial estate is north of Newcastle-under-Lyme. It has an ideal transport network and a good market. The site is not far from a supply of good skilled workers which are available. The site is a very pleasant working environment with trees and nice

surroundings. It has good power supplies and a market available nearby.

Examiner's comment

The candidate has chosen a local example and starts off well with clear reasons for its location. The answer then deteriorates into a series of general comments which do not explain its particular location. The answer does not make it clear what type of industry is found on the industrial estate.

Along with maps, graphs are the most popular type of resource used in the examination papers. Also, like maps, there are different types of graph which can be used in a variety of ways. You may be required to complete a bar graph, line graph, pie chart, divided bar graph, scatter graph or triangular graph, etc. You may have to describe or interpret information contained within the graph. Whatever the question, the key to dealing successfully with graphs is accuracy. If you are completing or taking information from any type of graph you must use the scale carefully and give a precise answer. Vague or incorrect measurements do not score marks.

The first two questions focus on a bar graph and are taken from a Foundation Tier paper.

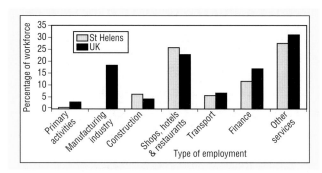

The questions show typical use of this kind of resource. The first question is a straightforward graph completion exercise.

5 **(a)** Complete the graph by using the following information:

25% of the workforce of St Helens is employed in manufacturing industry.   [1]

Remember the need for precision in drawing the bar to the correct percentage. Also remember that straight lines in the bar graph must be drawn with a ruler.

(b) How important is employment in manufacturing industry in St Helens compared with the UK as a whole? You should support your answer with figures.

[2]

This question is an example of graph interpretation. In other words you must use information from the graph to answer the question. Notice that in this question you only need to use the information about the manufacturing industry. To gain the second mark, figures must be read accurately from the graph.

Sometimes different graphs can be used to show the same information on the Foundation and Higher papers. The following line graph comes from a Foundation Tier question.

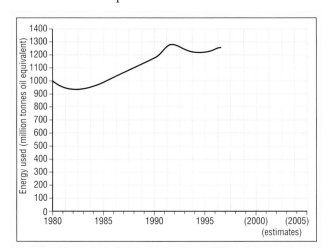

Compare that graph with the one below which comes from the Higher Tier paper.

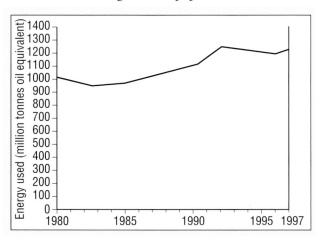

The graph on the Foundation Tier paper is made easier to use by the grid drawn as a background. This makes it easier to read off the values.

Now look at the difference in the questions between the two papers.

Foundation Tier question

6 (a) How much energy was used in 1990 ?

_____ million tonnes oil equivalent

[1]

Higher Tier question

7 (a) How much energy was used in 1997? [1]

As well as having to read the information accurately without the help of the squared background, the candidate on the Higher Tier paper also has to give the correct unit of measurement. Be careful not to lose the mark on this type of question by omitting the units (million tonnes oil equivalent), having done the task accurately.

Foundation Tier question

8 (a) Describe the changes in the amount of energy used between 1985 and 1995. Support your answer with figures and dates. [3]

Higher Tier question

9 (a) Describe the changes in the amount of energy used between 1980 and 1997. Support your answer with figures and dates. [4]

At first glance this question is almost identical on both papers. However, on closer inspection you will see that the time interval is greater on the Higher Tier and by looking at the graph you will see that there are more intricate changes as the line goes up and down. As with all graph interpretation questions, a good answer will be detailed and accurate as shown in the two examples below.

Foundation Tier answer

It has increased, the amount of energy used in 1985 was 1000 million tonnes oil equivalent whereas in 1995 it was 1240 million tonnes.

There was a definite increase in use between 1985 and 1990, then there was a terrific rise in 1991–1992, then a fall until 1995.

Examiner's comment

There were three marks allocated to this question:

1 mark: describe the general increase in use

1 mark: recognise that the increase is not constant

1 mark: use figures and dates from the graph to illustrate the answer.

This candidate fulfils all three demands of the question well.

Higher Tier answer

Between 1980 and 1982 there was a decrease in the amount of energy used, i.e. from roughly 1010 million tonnes to about 950 million tonnes of oil equivalent. Yet from mid 1982 to mid 1990 there was a steady increase in the amount used, i.e. from 950 to about 1100 million tonnes. Then from mid 1990 to 1992 there was a sudden increase of energy used from about 1100 million tonnes to about 1250 million tonnes. But from 1992 to 1996 there was a steady decrease in the amount of energy used, from about 1250 million tonnes in 1992 to about 1180 million tonnes in 1996. But from 1996 to 1997 there was an increase from 1180 million tonnes in 1996 to roughly 1210 million tonnes in 1997.

Examiner's comment

The four marks allocated to this question were equally divided between descriptive statements and supporting data drawn from the graph, i.e.

1 mark: description of 2/3 changes

2 marks: description of 4 changes

1 mark: use of 2/3 energy figures plus corresponding years

2 marks: use of 4 energy figures plus corresponding years.

This candidate describes the trends shown by the graph very methodically and illustrates each change with data. It is an excellent answer.

There are two common ways to show percentage information in graphs. These are illustrated below in a question which continues the focus on energy use.

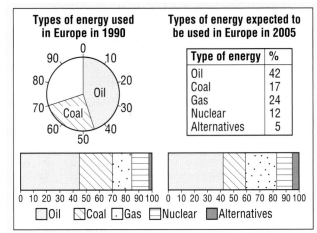

Type of energy	%
Oil	42
Coal	17
Gas	24
Nuclear	12
Alternatives	5

The two types of graph shown are a pie chart and a divided bar graph. The information given in the two sets of data is identical, just presented in different ways to meet the demands of the question. In the Foundation Tier paper the data for 2005 is presented as a table rather than another pie chart.

Foundation Tier questions

10 (a) (i) Complete the pie chart by using the following figures:

Type of Energy	Percentage
Gas	15
Nuclear	13
Alternatives	2

[2]

(ii) Which type of energy was used most in 1990? [1]

(iii) In 1990 what percentage of Europe's energy was obtained from coal? [1]

(iv) Describe what is likely to happen to the percentage of coal used between 1990 and 2005. [1]

(v) Which TWO types of energy are likely to increase in importance between 1990 and 2005? [1]

(vi) Suggest why the importance of different types of energy may change by 2005. [4]

Higher Tier questions

11 (a) (i) Describe the two main changes in the use of energy shown in the graphs.[2]

(ii) Suggest reasons why the importance of different types of energy may change by 2005. You should give at least three reasons and develop the points which you make. [5]

The questions reflect the different approach taken by the examiner in the Foundation and Higher Tier papers.

On the Foundation Tier paper	On the Higher Tier paper
Six sub-sections	Two sub-sections
Graph completion questions	No graph completion sections
Shorter, more specific graph and data interpretation questions	Longer, more open graph interpretation question
Fewer marks allocated to explanation	More marks allocated to explanation
No development of ideas required	Development of ideas is required

The final section is quite demanding and is testing the understanding of all candidates about likely changes in the use of energy in the next few years. One advantage of this type of question is that it deals with a topic where many different ideas can be discussed. The answers below show how different candidates answered this question on both the Foundation and Higher Tier papers.

Foundation Tier

Answer from candidate 1

This may happen as governments realise that oil and coal cause massive amounts of air pollution and that nuclear power is dangerous. Oil and coal may rise in price as they are non-renewable fuels.

Examiner's comment

The candidate understands that governments influence energy production and use but does not go on to say how government policy may affect energy use. The candidate also recognises two other issues associated with fossil fuels – pollution and cost.

Answer from candidate 2

People are more concerned with the environment each year. Also coal and oil resources are expected to run out in around 200 years so people are looking for a non-polluting, renewable source of energy.

Examiner's comment

The candidate deals with the two main issues concerning energy production – environmental pollution and exhaustion. The answer shows some understanding of the contrast between fossil fuels and alternative energy sources.

Answer from candidate 3

We are using different types of technology and so we don't really need coal, and we are trying to cut down on the pollution that we create.

Examiner's comment

The candidate does not expand the idea of new technology nor explain that pollution is connected to fossil fuels. There is potential for significant improvement in this answer.

Higher tier

Answer from candidate 1

The fossil fuels are running out. As they are non-renewable sources of energy, alternatives will have to be found and less usage of the stocks we have left will have to take place. Oil and coal are major contributors to the greenhouse effect as they are very large pollutants. People will want to use a 'cleaner' source of energy. Alternative energy will be more advanced by 2005. It will be more reliable. Once it is more widespread the equipment for it won't be as expensive to set up as it is now. LEDCs could take advantage of it.

Examiner's comment

Although not fluently expressed, the answer is concise and concentrates upon the changes from fossil fuels to alternative energy sources. Basic ideas are expressed, e.g. 'fossil fuels are running out', but there is also understanding shown of the more complex issue of pollution. The candidate also recognises that the development of alternative energy sources will continue which could reduce the cost of such schemes and make them more widespread throughout the world.

Answer from candidate 2

The importance of different types of energy may change by 2005 because science will develop the use of alternative sources like solar power, HEP and geothermal so that fossil fuels, which are rapidly diminishing in reserves, will become obsolete. Nuclear power will remain in a constant state of use because it will have been deemed to be safe by all the safeguards that modern technology provides.

Examiner's comment

This candidate recognises the change from fossil fuels to alternative energy sources, but does not expand upon these basic ideas. Reference is made to the nuclear debate but no details are given about how the process of producing nuclear energy will become safer.

Answer from candidate 3

As people become more concerned for the environment, highly polluting forms such as coal and oil will be reduced. Renewable sources will be more widely used. Nuclear fuel usage may increase as safety may be improved as time moves on. Coal supplies are running out as are oil sources and therefore, combined with government policy, the amount used will decrease.

Examiner's comment

The candidate fails to develop ideas and there is potential in this answer to gain more marks. Reference is made to pollution and the decline of fossil fuels but no other explanation is given for the likely increase in alternative energy sources. Fossil fuel exhaustion is also suggested but the candidate

fails to develop the important idea of how government policy may influence energy developments. If these ideas had been developed the answer could have scored maximum marks.

TASKS

1   The following are important topics in the People and Their Needs unit for which you need to know case studies. Make a check list of your case studies for these topics:
 • commercial farming system in an MEDC,
 • subsistence farming system in an LEDC,
 • rapid economic growth in an LEDC,
 • benefits and problems of tourism in either the UK or EU,
 • consequences of meeting the rising demand for energy,
 • consequences for a community in the UK of changes in energy production.

2   Here is another opportunity to become an examiner. You have seen two answers to the following Higher Tier question earlier in this chapter.

For a named and located example of an industry within the UK which you have studied, explain why the industry has developed there.   [7]

Below are two more answers to this question. Identify the main reasons why candidate 1 scored more marks than candidate 2.

Answer from candidate 1

Sunrise Strip is an area in the south of England near London in which there are many micro-electronics industries. These are footloose industries which have developed there due to the fact that they do not depend on raw materials, but instead their location is affected by the 'golf course effect' in which the operators are attracted to it due to the attractive environment of the Thames valley. They have also located here due to the proximity of good transport links such as Heathrow airport which links them to their market. Also, as they need a lot of new technology, they need to be near to government research centres and universities, (Oxford University is very close). These industries have also benefited from urban sprawl as this provides a large, skilled and educated workforce.

TASKS

Answer from candidate 2

Blakes of Loxley is located on the outskirts of Alcester because it is close to an industrial estate where people will be trained in certain aspects of work. It is also near some secondary schools and colleges which people are frequently leaving and looking for employment. The industry also developed at this location because it is close to a transport network which joins the whole country so it is a good location for transporting its refrigerated goods. The location is also near the centre of the UK so it is well placed for transportation to anywhere in the UK.

3 The two sets of graphs below show employment structure in a number of countries. You will be familiar with pie charts but may not have used a triangular graph before.

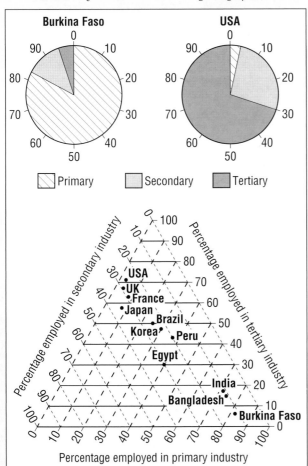

Percentage employed in primary industry

With a partner answer the following.

a What does employment structure mean?

b What unit is used to measure employment in both graphs?

c Use the table below to work out how to read data from a triangular graph. Copy out the table and fill in the gaps.

Country	Percentage employment		
	Primary	Secondary	Tertiary
Burkina Faso	82	12	6
	8	29	63
Bangladesh	74		15
USA			70

d Identify the differences between the employment structure of Burkina Faso and USA. Why do these differences exist?

4 Study the two scatter graphs below.

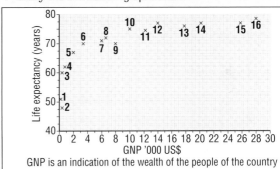

GNP is an indication of the wealth of the people of the country

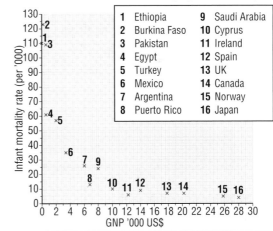

Answer the following questions which are taken from both the Foundation and Higher Tier papers.

a In which country is GNP $11000 and average life expectancy 75 years?

b In which country is the infant mortality rate highest?

c Describe the general relationship between GNP and average life expectancy.

d Explain why this relationship occurs.

e How is the relationship between GNP and infant mortality rate different from the relationship between GNP and average life expectancy?

f Explain why the relationship between GNP and the quality of life indicators is so different.

THE ENTRY LEVEL CERTIFICATE

The **Oral** accounts for 20% of the marks for the Certificate of Achievement. Near the end of the course, you will be given some information on one of the topics you have studied. It could be a map or photo or a graph or a newspaper article. Your teacher will ask you some questions about the topic and record the Oral on tape.

Your teacher will arrange a time to meet and will find a room – possibly your classroom or an office – to record your interview on tape. No other pupils will be there, so you will be able to show what you know and understand. There is nothing to write down. Just talk about the resources you are looking at and be prepared to use your knowledge of real examples (**case studies**).

Here is an example of a case study which you may need to use in an Oral.

People and their needs – Industry

CASE STUDY: Changing jobs in Stoke-on-Trent

Jobs are changing today in towns and cities across the country. Stoke on Trent, as well as having many jobs in pottery factories (**secondary**), used to have many jobs in coal-mining (**primary**). Hem Heath coal mine was one of the most efficient mines in the country. Photo A shows the entrance to the mine today – it has closed and there are now no working coal mines in the area. Many people still work in the pottery industry making cups, plates and ornaments. There are a number of other secondary jobs on **industrial estates** too. Photo B shows a factory on Newstead Industrial Estate on the edge of the city. Across the road is a new **business park**, Trentham Lakes. Photo C shows the entrance and the new Worldgate offices. All this is on the land that used to be Hem Heath coal mine. In just a few years, the types of jobs for people in this area of Stoke-on-Trent have changed a great deal.

TASKS

1   What happened to Hem Heath coal mine?

2   What changes do the photos show for primary and secondary jobs?

3   a   What are industrial estates?

  b   What are the advantages of having many factories all in one area ?

Fieldwork Tip – As part of your coursework, you could find out about a business park or industrial estate near you. Map it. Make a note of all the factories or offices on the site and find out what they do. Describe its location and work out how many offer jobs in primary, secondary or tertiary work.

PHOTO A

PHOTO B

Oral Task

a) Introduction

▶ General conversation with questions about work.
Have you got a part-time job?
What do you have to do ?

Where are you going on work experience?
What will you have to do?

What job do you want to do in the future ?
What will you have to do?

▶ Look at the three photos. They show three types of work.
One of them was primary work. Which photo is it?

Can you tell me how these three types of job differ?

(Can the student identify primary, secondary and tertiary activities?)

b) Describing and explaining

▶ Referring to the local area.
Can you name three jobs in our local area from each sector of industry?

Can you name a factory or other place of work for each example you give?

▶ Imagine you were going to open a factory making _____ *(use local example)*
What things would you need to make the factory successful?
(Can the student identify the factors of production – labour etc?)

PHOTO C

c) Case Study

▶ Referring to the local area.
Name a local factory.

What do they make?

Why do you think the factory was located where it is?

If you were setting up a new factory near here, where would be the best place for it?

Why would you locate it there?

How is that location better than the other factory you were talking about?

▶ The best Orals end up more like a conversation, than a question and answer session. Your teacher will have some questions but they are only a 'guide' to help get the best out of you.

People and the Environment

LIMESTONE

The extraction of raw materials by mining or quarrying

Carboniferous limestone is a hard grey rock that often has deposits of other minerals and ores within it, for example, lead and fluorspar. It is resistant to weathering and forms upland areas which have a distinctive scenery. Areas such as Castleton (page 230) and Dovedale in the Peak District or Malham in North Yorkshire attract many visitors each year. Many limestone areas are located in national parks (Figures 4.1 and 4.3). These rural areas have few industries and limestone quarrying provides employment.

Limestone is an important raw material and is used for construction, road stone and to make cement (Figure 4.2). The demand for limestone depends on the construction industry. If demand for new houses or infrastructure projects is low, production falls as, for example, less cement is needed.

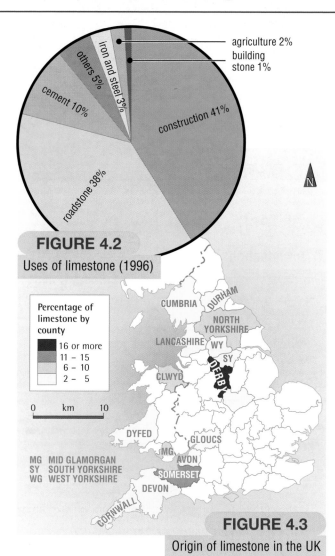

FIGURE 4.2

Uses of limestone (1996)

agriculture 2%
building stone 1%
iron and steel 3%
others 5%
cement 10%
construction 41%
roadstone 38%

Percentage of limestone by county

16 or more
11 – 15
6 – 10
2 – 5

0   km   10

MG MID GLAMORGAN
SY SOUTH YORKSHIRE
WG WEST YORKSHIRE

FIGURE 4.3

Origin of limestone in the UK

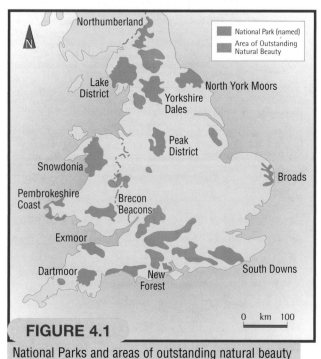

Northumberland

National Park (named)
Area of Outstanding Natural Beauty

Lake District
North York Moors
Yorkshire Dales
Peak District
Snowdonia
Broads
Pembrokeshire Coast
Brecon Beacons
Exmoor
Dartmoor
New Forest
South Downs

0   km   100

FIGURE 4.1

National Parks and areas of outstanding natural beauty

Quarrying in Derbyshire

The largest area of limestone quarrying in the UK is found in Derbyshire. Several quarries are within the Peak District National Park. Many were operating before the National Park was established. Although the aims of the 1949 National Park and Access to the Countryside Act included the conservation of natural areas and wildlife, it also included supporting the economic and social needs of local communities living in national parks. Achieving the first often caused conflict with the second.

While there is legislation to protect new areas from quarrying, many Peak District quarries had mineral rights which pre-dated the National Park. The Peak District boundary was drawn to skirt around many of the older quarries, for example, to the east of Buxton.

There has been a decline in employment in mineral and metal extraction in Derbyshire, but it still accounts for 10% of male employment, especially in the rural areas where unemployment is 8% (Figures 4.4 and 4.5).

	Primary	Secondary	Tertiary
Derby City	5	32	63
Derbyshire Dales	17	16	67
High Peak	15	24	61
SE Derbyshire	25	17	58
County average	11	28	61

Figure 4.4 Employment in Derbyshire (%)

	1981	1991	1996
Amber Valley	14	12	10
Bolsover	46	29	8
Chesterfield	18	12	2
Derby City	5	5	2
Derbyshire Dales	11	12	12
Erewash	5	1	1
High Peak	19	14	13
NE Derbyshire	21	8	2
SE Derbyshire	29	22	14

Figure 4.5 Mineral and metal extraction employment, 1981–96 (%)

Environmental impact

The extraction of limestone has a massive impact on the environment (Figure 4.6). Earth mounds are used to reduce the noise of explosions and of machinery. Water sprays are used to reduce the spread of dust. However, quarrying can still be heard over a wide area. The quarries form large, white scars on the landscape. Trains are used to move some of the limestone, but much is transported by lorries, causing problems as they pass down narrow roads and through villages already busy with the tourist traffic. The heavy lorries stress the road surface and contribute to noise and air pollution.

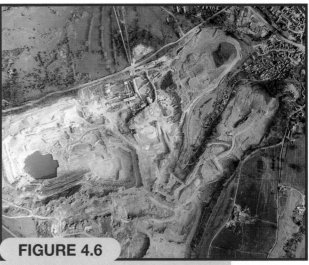

FIGURE 4.6
Limestone Quarry, Wirksworth, Derbyshire

In recent years there have been several applications to extend quarrying on privately owned land in the Peak District. The legislation is complex, but local authorities use prohibition orders to control developments. An exception is the £23 million scheme to strengthen the dam for Ladybower Reservoir, using 400 000 tonnes of limestone quarried from Win Hill. Local conservation groups are working with the construction company to minimise damage and ensure restoration.

TASKS

1 Which National Parks in the UK are affected by limestone quarrying?

2 Suggest why there is a link between limestone production and construction.

3 Use Figure 4.4 to draw a bar graph comparing Derby City with High Peak and Derbyshire Dales. Describe what the bar graph shows.

4 Using the photographs and map extract (page 20), describe the impact of limestone quarrying on the landscape.

5 What are the benefits and problems caused by limestone quarrying to people living in nearby settlements?

6 Use Figure 4.6 to draw a plan of the area shown. Label the following features:
  a the stock pile of limestone, the conveyor belt, the buildings/grinding plant, lake, quarry faces being worked, the direction the quarry is likely to expand, an area of former working,
  b the settlement of Wirksworth, the main roads, the disused railway line.
  c Estimate the area of the quarry.

7 If a proposal to expand this quarry were submitted, do you think that it should be accepted or refused? Justify your answer.

TROPICAL RAINFORESTS

FIGURE 4.7

Distribution of tropical rainforests

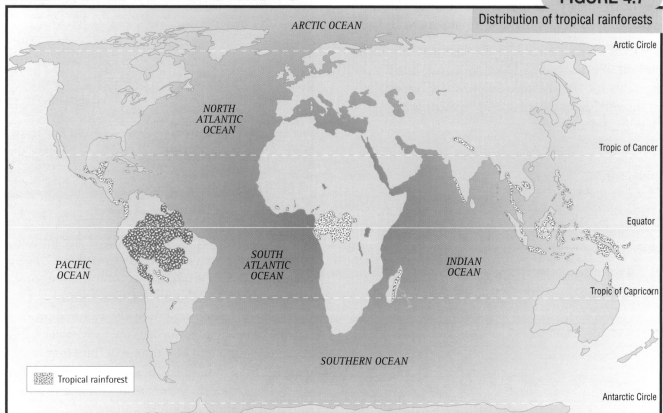

Tropical rainforests are the most productive ecosystem in the world. They contain examples of over half the world's species of plants and animals, over one-third of the world's trees grow there, and they have a vital role to play in world climate and the content of the atmosphere. Figure 4.7 shows the distribution of the world's rainforests. They lie in a belt along the Equator and within the tropics. Although there is some rainforest in 70 countries, three countries between them account for more than half the total area. The **ecosystem** has evolved over millions of years with a number of key elements working together to create this unique environment.

The climate graph (Figure 4.8) is for Manaus, which is located in middle of the Amazon Basin in Brazil. It has a typical equatorial climate – hot, wet and humid all year round. It is unusual in that it has no seasons. This climate encourages rapid growth of vegetation. Temperatures are high throughout the year. Evening temperatures rarely fall below 22°C. Daytime temperatures rise up to 32°C. The annual range of temperature is very small (2°C). The pattern of weather is the same every day – early morning sunrise, high temperature by mid-morning evaporating the previous day's rainfall, trapped on the dense vegetation; by mid-afternoon the rising air currents cool, leading to condensation, the formation of billowing cumulonimbus clouds, and ultimately heavy rain and often thunderstorms. This is **convectional rainfall**.

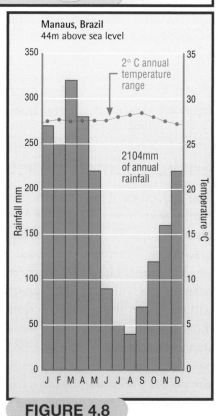

FIGURE 4.8

Climate graph for Manaus

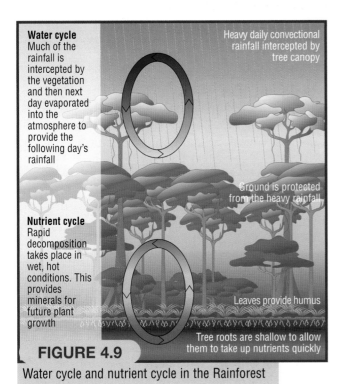

Water cycle Much of the rainfall is intercepted by the vegetation and then next day evaporated into the atmosphere to provide the following day's rainfall

Heavy daily convectional rainfall intercepted by tree canopy

Ground is protected from the heavy rainfall

Nutrient cycle Rapid decomposition takes place in wet, hot conditions. This provides minerals for future plant growth

Leaves provide humus

Tree roots are shallow to allow them to take up nutrients quickly

FIGURE 4.9

Water cycle and nutrient cycle in the Rainforest

A high proportion of this rain never reaches the ground. The dense vegetation cover intercepts much of it, protecting the soil from erosion. The roots of the trees also bind the soil together. The rainforest soils, although deep, tend to be very infertile. The hot, humid climate encourages rapid decomposition of organic matter by fungi and micro-organisms. Thus the rainforests generate their own mineral supply, as a rich humus layer builds up at ground level. There are two cycles occurring in the rainforest – a water cycle and nutrient cycle. These are shown in Figure 4.9.

At first glance, rainforest vegetation with its vast tangle of climbing plants and creepers appears chaotic. There is, however, a structured order to the forest. The vegetation has adapted to the local conditions. For example, leaves have evolved drip tips to get rid of the heavy raindrops and large buttress roots stand above the ground to give the tall, shallow rooted trees support.

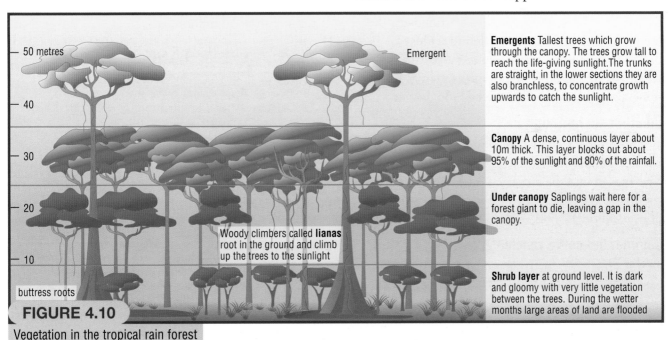

50 metres

Emergent

40

Emergents Tallest trees which grow through the canopy. The trees grow tall to reach the life-giving sunlight. The trunks are straight, in the lower sections they are also branchless, to concentrate growth upwards to catch the sunlight.

30

Canopy A dense, continuous layer about 10m thick. This layer blocks out about 95% of the sunlight and 80% of the rainfall.

20

Under canopy Saplings wait here for a forest giant to die, leaving a gap in the canopy.

Woody climbers called **lianas** root in the ground and climb up the trees to the sunlight

10

buttress roots

Shrub layer at ground level. It is dark and gloomy with very little vegetation between the trees. During the wetter months large areas of land are flooded

FIGURE 4.10

Vegetation in the tropical rain forest

TASKS

1 Describe the distribution of the world's rainforests.

2 Study the climate graph for Manaus. Describe and explain its characteristic features.

3 You can discover more about how the rainforest ecosystem works from CD-ROMs and the Internet. You could use the information from these sources together with the pages in this book, to create a word processed or desktop published

report about the Rainforest Ecosystem

Useful websites: Rainforest Action Network
http://www.ran.org/ – Rainforest Information and the Kids Corner includes information about rainforests.

Rainforest Alliance includes information about the importance of rainforests
http://www.rainforest-alliance.org/

DEVELOPMENT PROJECTS IN THE AMAZON

The reasons why Amazonia is being deforested are mostly linked to the economic development of the countries it covers, particularly Brazil. Rapid population growth since the 1960s meant that Brazil needed more land for people to live on and to farm. There were also political attractions for the government, who gained international prestige from big new schemes. Developing Amazonia took people's minds off national issues such as poverty and landlessness. Exploiting rainforest resources was also seen as a way of reducing the country's huge national debt.

Transport

The first requirement to develop the rainforest was to make it accessible. Over 12 000 km of new roads have been built in the rainforest, starting with the 5300 km Trans-Amazonian Highway. This provided the means for people to move into the region, as well as allowing resources such as timber, minerals and farm produce to be brought out.

Small-scale farming

The government offered poor rural people in Brazil plots of land in the rainforest. Thousands of families have moved into the area from other areas of the country to take advantage of this scheme, particularly from the drought-ridden north-east region of Brazil.

Commercial cattle ranching

This is usually run by large trans-national companies. Ranchers burn areas of forest, replacing trees with grass. Large-scale cattle ranches occupy 25% of the cleared areas of rainforest in the Amazon.

Forestry

Tropical hardwoods such as ebony and mahogany are obtained by logging companies. The timber is a valuable source of foreign income for Brazil, but little attempt is made to replant. There is a large demand for these woods from MEDCs.

Minerals

The rainforest possesses great mineral wealth, including iron ore, bauxite, manganese, diamonds, silver and gold. Mining companies have cut down trees and built roads and railways through the forest to get to these deposits. Carajas contains the largest single iron ore reserve in the world – 18 billion tonnes. Miners will be busy here for more than 300 years. The European Union and the World Bank provided a large share of the financial backing for the project. The short-term gain for the EU was a 15-year guarantee of 13.6 million tonnes a year at a bargain price.

Hydro-electric power

The rainforest has an unlimited supply of water, and good conditions for HEP development. More than 125 HEP dams are to be built in the next 15 years, but vast areas of forest will be flooded by the creation of large lakes.

Settlement

Large areas of forest have been cleared for the development of new settlements. In 1960, the population of the Amazonia was 2 million. Now it is more than 30 million.

TASKS

1 Describe and explain the ways that the Amazon rainforest is being used to create wealth for Brazil and improve the quality of life for its people.

2 The Brazilian government's point of view about developments in the rainforest is provided on the Brazil Embassy website: http://www.brazil.org.uk
   The following is a quote taken from this website.
   'It would be unreasonable to expect a developing country in Brazil's situation, with pressing social needs and so many natural resources awaiting exploration, research and protection, to consider its own development as unattainable due to environmental considerations. Development, as it has been experienced since the Industrial Revolution, has entailed environmental costs. The environmental record of the industrialized countries themselves bear this point out clearly.'
   Explain the point of view being expressed.
   If you have access to the Internet, visit this website to discover further information about the Brazilian Government's viewpoint attitude to developing the rainforest.

3 Until recently the Brazilian government has monitored levels of deforestation using satellite images. Every 16 days, the LANDSAT 5 satellite sends 229 images of the Amazon Basin back to earth. What are the advantages of using this method to monitor environmental change in the rainforest? If you have access to the Internet, you can visit the satellite monitoring project at its website: http://www.dpi.inpe.brgrid/quick-looks
   Try to obtain further details about the project.

FIGURE 4.11

Carajas – the world's largest iron ore mine

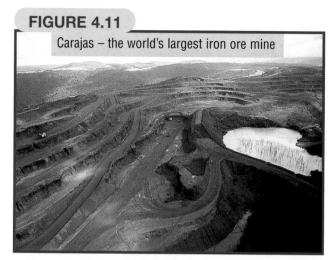

FIGURE 4.12

Cattle ranching in Para State, Brazil

FIGURE 4.14

Itaipu hydroelectric dam

FIGURE 4.13

New road through partly deforested land, Amazonia

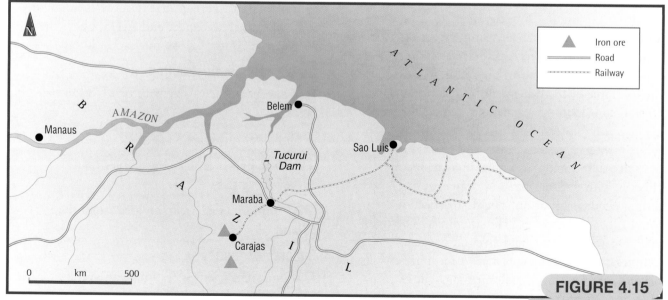

ATLANTIC OCEAN

Iron ore
Road
Railway

B
R
A
Z
I
L

AMAZON

Manaus

Belem

Tucurui Dam

Sao Luis

Maraba

Carajas

0   km   500

FIGURE 4.15

The location of Carajas

ENVIRONMENTAL CONSEQUENCES

Cleared forest – heavy rainfall hits the ground

Soil erosion and flooding caused by increased run-off

Fewer trees to intercept rainfall, therefore rates of evapo-transpiration decrease. Long term the levels of rainfall decrease as the water cycle is broken

Nutrients in soil washed downwards (leaching). Soil becomes less fertile

No tree roots to hold soil together. No fallen leaves to provide humus

The nutrient cycle is broken with the large scale felling of trees. The environment is no longer self-sustaining

FIGURE 4.16

The consequences of deforestation

Economic gains made by countries such as Brazil when developing the rainforest need to be offset against the environmental consequences of deforestation. Some of these are shown in Figure 4.16.

Rainforests cover less than 2% of the earth's surface, yet they are home to 40 to 50% of all life forms on our planet – as many as 30 million species of plants, animals and insects. The rainforest is a finely balanced ecosystem. Once the trees are cut down, all other elements of the environment are affected. The animals and insects lose their habitats. Without the forest canopy, the soil is exposed to the full force of the heavy rainfall, leading to soil erosion and flooding. In the long term, the absence of the canopy interrupts the daily rhythm of rainfall. Interception decreases and there is less evapotranspiration. This results in a change in climate, with reduced levels of rainfall and ultimately desertification. The destruction of the forest poses a huge global climate threat. The trees take in carbon dioxide and give out oxygen. Deforestation affects the planet's atmospheric balance, accelerating global warming. The burning of large areas of trees further pollutes the atmosphere, increasing levels of carbon dioxide.

Environmentalists fear the loss of the rainforest's unique biodiversity. They point out the Amazon is home to plant species which provide everything from chocolate to today's most important medicines. They warn that if the area is destroyed, its untold secrets will never be revealed. Over half of our modern medicines come from the rainforests, including painkillers and quinine, used to treat malaria. Perhaps the rainforests hold cures to diseases such as cancer and AIDS, but if the destruction continues, we may never know.

Global rates of destruction

1 ha per second: equivalent to two US football fields

= 60 ha per minute

= 86 000 ha per day: an area larger than New York City

= 31 million ha per year: an area larger than Poland

Species extinction

Distinguished scientists estimate an average of 137 species of life forms are driven into extinction every day; or 50 000 each year.

Wednesday, April 14, 1999 Published at 20:36 GMT 21:36 UK   BBC Online News

Amazon forest loss estimates double

Logging's damage is less obvious than forest clearance, but no less real

By Environment Correspondent, Alex Kirby

The true extent of rainforest damage in the Amazon is more than twice as great as present estimates suggest, researchers say. Daniel Nepstad: 'The degradation will continue'. The team says field surveys of logging and burning show far more deforestation than satellite monitoring has revealed.

The researchers interviewed 1393 wood mill operators, representing more than half the mills in 75 Amazonian logging centres. As well, they interviewed 202 landlords, whose properties covered 9200 sq. km.

They found that logging crews annually cause severe damage to between 10 000 and 15 000 sq. km of forest that are not included in current deforestation estimates.

Insidious damage

They also discovered that fires burning on the surface consume large areas of forest which again are not recorded. The researchers say the failure so far to register the much greater loss rate they have discovered is because the loggers reduce tree cover, but do not eliminate it. By contrast, ranchers and farmers deforest land in preparation for pasture and crops by clear-cutting it, and by burning whole areas. The more the forest burns, the more vulnerable to fire it becomes. And where logging and fires have caused damage, they say, the vegetation will grow back fast enough to dupe a satellite.

The only reliable way to find out what is happening is by field surveys. Logging and surface fires seldom kill all the trees. But they help to make them more vulnerable. Logging increases the flammability of the forest by reducing leaf canopy coverage by up to 50 per cent. This lets the sunlight strike through to the forest floor, where it dries out the organic debris created by the logging.

Satellites not enough

But they found that only about a tenth of the area classified as forest actually supported undisturbed forest. The researchers say: 'Satellite-based deforestation monitoring is an essential tool in studies of human effects on tropical forests, because it documents the most extreme form of land use, over large areas, and at low cost.' But this monitoring needs to be expanded to include forests affected by logging and surface fire if it is to accurately reflect the full magnitude of human influences on tropical forests.

TASKS

1 Why do environmentalists think tropical rainforests are so important?

2 Describe and explain how deforestation affects the environment at a local and global scale.

3 Read the news article from BBC Online News, outlining the results of new research about the impact of logging on the rainforest.

  a How was the research conducted?

  b Why does logging cause greater damage than previously thought?

  c Why do the researchers feel that the Brazilian government's monitoring of deforestation using satellite images alone is not accurate enough?

4 Environmental pressure groups are mounting worldwide campaigns to raise awareness of the plight of tropical rainforests. Several of these groups use the Internet as part of their campaign. If you have access to the Internet, research the following sites, and produce a word processed report, outlining the viewpoint of environmentalists about deforestation.

Rainforest Action Network
http://www.ran.org/
Rainforest Alliance
http://www.rainforest-alliance.org/

5 The United States Geological Survey's website, Earthshots, investigates environmental change through the use of satellite imagery. One of its case studies examines the impact of deforestation in the Rondônia area of the Amazon rainforest. If you have access to the Internet, visit this case study at the following web address:

http://www.usgs.gov/Earthshots

Download the satellite images of the case study area in Rondônia for 1975 and 1992, copy and paste the images into a desktop publishing program and, using the software tools, label the changes that have occurred in the area.

STEWARDSHIP AND SUSTAINABLE DEVELOPMENT

There are two conflicting forces at work in the rainforest, economic and environmental. Economic exploitation of the rainforest by LEDCs that need to generate industrial growth, to trade on the world market and to improve the quality of life of their people has environmental consequences. Environmentalists feel that deforestation will have major implications for the whole planet. As a result, some development is inappropriate, because in the long term, it will stop people from enjoying a better quality of life. A more appropriate way to use the rainforest is sustainable development. What are needed are schemes that use but do not waste resources. Rather than cut the rainforest down, better uses could be made of the existing stands of trees. Sustainable development requires good planning and international cooperation, together with a commitment to conservation.

The attitude of the country's government is most important. The democratically elected governments of Brazil in the 1990s have been much more conservationist than the previous military governments. A wide range of sustainable and effective initiatives is needed to keep the forest alive while improving the quality of life for the 30 million people who live there. Obviously this can only be achieved if economic alternatives and solutions to destructive logging can be found. Governments and environmental groups have a number of sustainable schemes.

FIGURE 4.17

Harvesting the rainforest, Brazil

Rubber tapping

This is a traditional and ecological activity that has been part of the life of Amazonia since the beginning of the twentieth century. The activity does not damage the forest as it is not necessary to cut down a rubber tree to extract the latex. Today about 63 000 families of rubber tappers live in extractive reserves. These reserves were established by the Brazilian government in order to allow rubber tappers to carry out their traditional activities. They cover approximately 1% of the Amazon rainforest. The National Council of Rubber Tappers would like about 10% of the Brazilian Amazon to be extractive reserves. Presently, only 5000 tonnes of rubber are produced from the Amazon, equivalent to 1.4% of the national market for rubber in Brazil.

Non-timber forest products

The economic value of a forest extends beyond the value of its timber. Traditionally, tribal people have harvested many forms of produce from the forest without destroying its ecosystem. Fibres, fruits, seeds, flowers, nuts and honey are just some of the many non-timber forest products from the forest. The fruits of the Acai Palm, for example, are used to make a wine rich in minerals. One palm tree produces about 20 kg of fruits per year. The tasty, dark violet wine is the most important non-wood forest product in terms of money from the river delta of the Amazon. In 1995, almost 106 000 tonnes were produced at a value of US $40 million. The camu-camu contains larger concentrations of vitamin C than any other fruit known today and is imported to the USA to produce vitamin tablets.

Additionally, many Indian tribes traditionally collect Brazil nuts and, for many, this is their main source of income.

Medicinal plants

More than two-thirds of all mass-produced drugs are derived from medicinal plants. At present, there is trade in about 650 species from Amazonia.

Ecotourism

This has a huge potential in Amazonia but it is poorly developed. Ecotourism would guarantee low environmental impact on the rainforest. Environmentally sensitive accommodation can be provided for visitors. Ecotourism could also guarantee that wealth creation from such activities would directly benefit local communities, who would provide guides and maintain the accommodation.

With its natural beauty, Amazonia offers many different options for ecotourism and adventure tourism such as trekking, rafting, diving, cruising, birdwatching and wildlife observation. In 1997, there were 16 jungle lodges in the Amazon registered in the Tourist Office of the Amazonas state, offering 1007 beds in total. Further development has to be carefully monitored in order to ensure the sustainability of the expansion of this industry. Figure 4.19 on page 190 is the mission statement of a major ecotourism company operating in the Amazon.

Sustainable logging

The Amazon rainforest is the greatest reserve of commercial timber in the whole world, estimated at around 60 billion cubic metres. At the moment, less than 1% is logged sustainably. Ecologically responsible forest management seeks to ensure that the ecosystem of the forest is not damaged and low volumes are extracted (see the news article on page 191). The impacts on the plant and animal life in the forest from this kind of logging are very small indeed.

FIGURE 4.18

Panning for gold, Venezuela

'Lagamar Expeditions is a company specializing in adventure travel in the environmentally rich areas of the Amazon region of Brazil. Our mission is to help spread the understanding of our environment and importance of our environment's biodiversity, by sharing secrets and mysteries surrounding us and our world. We are committed to rainforest conservation and furthering the empowerment of the peoples of the Amazon to become full participants in natural resource decision making.

Amazon • Pantanal • Galapagos • Paddling Programs • Brazil • Ecuador • Peru • Chile • Costa Rica

RECOMMENDED BY
BRITANNICA®
INTERNET GUIDE
BY ENCYCLOPÆDIA BRITANNICA®
www.ebig.com

Safari Guide

Multi Media

The Habitat

Booking Info

About Us

Books

Exploring the Amazon & Pantanal regions of South America

We outfit special interest expeditions for birdwatchers, botanist, photographers, etc

BEST OF
Buy IT
OnLine

Rare Trips for the Adventurous

**Got a burning desire to explore Amazon's deepest jungles,
trek the Inca Trail to magical Machu Picchu,
paddle the wild waters of Chile's Rio Futalehfu,
participate on a Photo Safari in the Amazon & Pantanal?**

- Then you've landed at the right place -

**Lagamar Expeditions provides responsible adventure tourism in the
Amazon, Pantanal, and other regions of Latin America.
See our Safari Guide for program details.**

For further program information please call: 1-800-823-8531.
E-mail address: WebMaster@lagamar.com

Lagamar Expeditions' objective is to provide sustainable tourism in natural areas that interpret the local environment and culture, further the tourist's understanding of them, foster conservation, and add to the well-being of the local people. Tourism is based on a guest/host relationship. It is vital for the guest (tourist) to respect the ownership, rights, and wishes of local people and communities, tread lightly on the environment, and contribute to the local economy rather than exploit it.'

The 'mission statement' of Lagamar Expeditions, as shown in their website

FIGURE 4.19

Lagamar Expeditions website

BBC Online News   Monday, April 19, 1999 Published at 12:38 GMT 13:38 UK

Amazon logging deal agreed

An area the size of Belgium was cleared in 1998

Brazil has lifted a ban on new logging permits in the Amazon Rainforest after landowners and loggers agreed to slow their rates of forest destruction. However, the country's environment ministry says any new permits for felling trees will be subject to strict guidelines. Some conservationists, who doubt whether any new controls can be effectively enforced, have questioned the wisdom of issuing permits. Brazil's authorities refused to issue new logging permits in February after they discovered more than 15 500 square kilometres (6 000 square miles) – an area the size of Belgium – was cleared in 1998. This was a rise of nearly a third on the previous year's clearing rates.

New commitment

The new guidelines limit local farmers to clearing just three hectares a year. Logging firms signing up to the agreement have also pledged to make better use of areas which have been partially logged, and to limit their use of fire in clearing forest. Antonio Prado, a spokesman for IBAMA – Brazil's environment protection agency – says the agreement marks the first time such a commitment has been reached between government, loggers and environmental groups.

Mr Prado said convincing Brazilian loggers to accept sustainable practices will also require financial incentives, and he said will be seeking economic support from Western countries to help IBAMA enforce its new rainforest guidelines.

Effective enforcement

The environmental campaign group, Friends of the Earth, says the agreement is a welcome step. But its spokesman, Tony Juniper, believes IBAMA will have to overcome a history of failing to enforce conservation laws if the rainforest agreement is to succeed. If the new measures succeed, they could make a significant difference in preserving one of the world's most important natural resources.

BBC Online News

*Friday, December 5, 1997
Published at 17:48 GMT*

Sue Branford

Eduardo Martins, head of Brazil's main environmental body, IBAMA, has announced a new plan to conserve the Amazon rainforest. Mr Martins, who is in London with the Brazilian president, Fernando Henrique Cardoso, said Brazil had decided to accept the request, made recently by Prince Philip, the president of the World Wide Fund for Nature, that every country in the world should conserve 10% of its forest cover.

The announcement, made in London by Eduardo Martins, took environmentalists by surprise. Mr Martins said that the Brazilian government had decided to respond positively to the appeal made by Prince Philip.

Brazil, he said, would increase the percentage of its forest that is protected from the current level, of less than 4%, to 10%. The new commitment will mean that the government will create new ecological reserves, covering four million hectares. Mr Martins said that just as important was another new measure. The government, he said, will be creating wide ecological corridors between all the existing reserves. These will allow birds and animals to move freely between the protected areas, greatly increasing their efficiency in protecting biodiversity. These conservation measures are to be paid for by the Group of Seven industrialised countries, who are giving Brazil $250m for Amazon conservation.

TASKS

1. Write a definition of sustainable development.

2. Explain why sustainable development is important to the future of the world's rainforests.

3. Read the examples of sustainable development projects. For each example, describe the scheme and explain why it is sustainable.

4. Read the news article about logging agreements in Brazil.
   a. Why had the Brazilian government stopped issuing logging permits?
   b. Why has the government started to issue them again?
   c. How can loggers be persuaded to use sustainable methods?

5. Why are the Brazilian government prepared to conserve areas of rainforest?

6. Figure 4.19 is an Internet page for a travel company called Lagamar, which specialises in adventure holidays in the rainforest. Read the mission statement of the company and explain how ecotourism can contribute to sustainable development. Visit the company's website to find out more:
   http://www.Lagamar.co./index.html

7. Many environmental pressure groups have internet sites to help promote their campaigns. Visit the websites below and find out more about sustainable development projects. You could design a desktop published flyer outlining projects researched from the websites.
   Greenpeace http://www.greenpeace.org/
   Rainforest Action Network http://www.ran.org/
   Rainforest Alliance http://rainforest-alliance.org/

LAND USE CONFLICTS IN NATIONAL PARKS

Although 75% of the UK's population now lives in towns and cities, most people enjoy the countryside – but are we loving it to death?

In the 1950s, national parks were set up to:

▶ preserve and enhance the natural beauty, wildlife and cultural heritage of the area

▶ promote the understanding and enjoyment of the countryside by the public

▶ foster the economic and social well being of local communities within the national park.

FIGURE 4.20

Wild ponies in Dartmoor National Park

Ten large areas of beautiful and wild countryside in England and Wales were originally chosen. The Broads, New Forest and South Downs were added more recently and now Scotland's first national parks are being created. Each has its own National Park Authority to run it, but, in total these authorities own only about 3% of the national parks. Two-thirds of the land is privately owned by farmers, the rest by organisations such as the National Trust, Forestry Commission, the water companies and the Ministry of Defence.

The main challenge facing national park authorities is planning for more visitors. Already the national parks receive more than 100 million 'visitor days' a year. People have

▶ more money to spend on leisure

▶ more time for holidays, especially short breaks

▶ better transport links from the cities.

Visitors tend to forget the national parks are home to 300 000 people. At busy times, there is traffic congestion and not enough parking. Visitors expect toilets, picnic areas, litter collection and tourist shops. In some national parks, house prices are high because rich people from cities have paid more for their second home than most local people can afford. Some villages have many retired people who have moved in, which changes the community.

TASKS

1 Why were national parks set up?

2 Why are visitor numbers increasing?

3 In the USA, national parks such as Yellowstone are government owned. How does that make it easier to run than a national park in the UK?

4 Many communities in national parks
   a have empty homes for much of the year,
   b are short of young people
   c have many old people.

   Suggest reasons for each of these.

You can find out more information on the Council for National Parks' website: http://www.cnp.org.uk

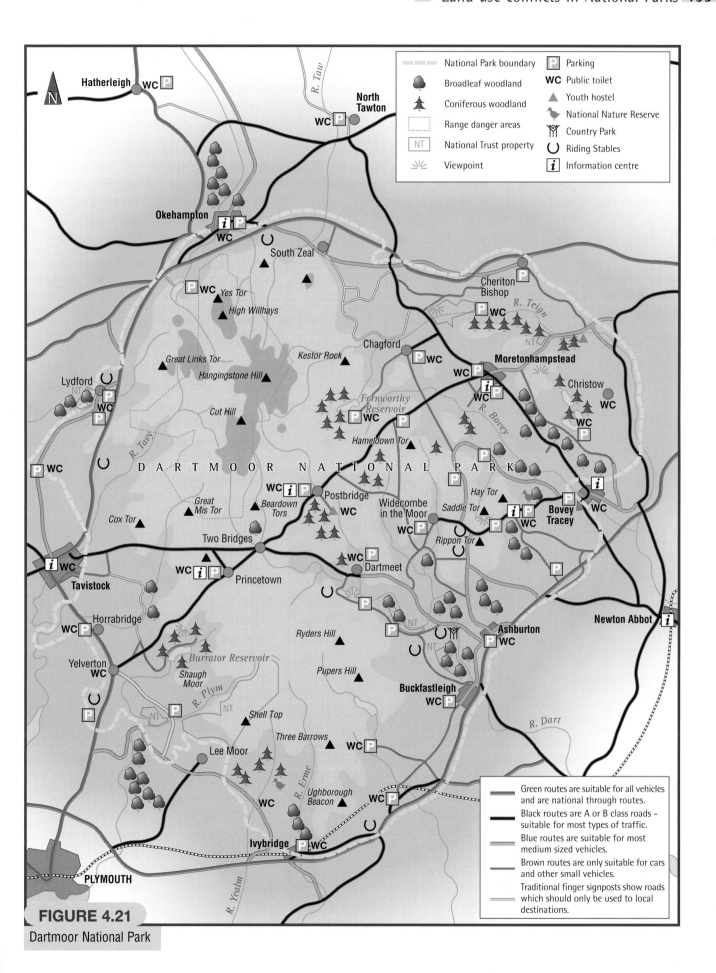

FIGURE 4.21

Dartmoor National Park

DARTMOOR

Erosion

Dartmoor is the largest and wildest area of open country in the south of England. Two-thirds is granite moorland, with tors which provide spectacular views. The most accessible are **honeypots** to visitors. Hay Tor, for example, is only 300 m from a car park (Figure 4.23). Too many feet trample the vegetation here. Heavy rain has washed away the exposed soil, creating gullies. In some areas, vehicles have been driven onto open land, causing more erosion, and horseriding has damaged the historic granite tramway at Haytor.

The National Park Authority's strategy is to turf or re-seed the most eroded parts. A new path at Haytor has been created to spread the load and temporary fences used to divert horses and walkers away from the eroded areas. Grass banks and blocks of granite are used to keep vehicles off open land.

Military training

14% of Dartmoor is used by the Ministry of Defence for training. Large areas of high moorland in North Dartmoor are closed to the public when live ammunition is in use and there has been damage to ancient monuments and wildlife. The area used is now only about half of what the army had when Dartmoor became a national park. The authority would like to see the end of all military training on the moor.

There is opposition to any more reservoirs. The last battle was won in 1970, when the site for a new reservoir to supply Plymouth was chosen at Roadford outside the national park.

Water supply

Dartmoor has no natural lakes but there are eight reservoirs, such as Burrator Reservoir (Figure 4.22) where valleys have been flooded to store water for Devon's towns and cities. Here, summer water shortages are common. Rainfall is high on the moor, where the wettest parts average more than 2 000 mm a year. (For comparison, London receives 610 mm.)

China clay

The granite that forms the high moorland and the tors also weathers into kaolin or china clay. It is used to make paper shiny. Mining it is a major industry on Dartmoor and china clay is exported all over the world. There are vast reserves in the south of the national park around Lee Moor. Here is one of the largest china clay pits in the world, over 90 m deep. However, vast quantities of waste are produced and dumped in large heaps near the open-cast mine. Important areas for recreation and wildlife have been under threat.

Shaugh Moor was saved when a public inquiry persuaded the china clay company to alter its plans. Lee Moor's waste heaps are being landscaped but it will be many years before they look like natural hills.

FIGURE 4.22

Burrator Reservoir

FIGURE 4.23

National Park Warden picking up litter at Haytor car park

Conserving the environment

Dartmoor has two large areas of blanket bog. With very heavy rainfall and poor drainage, the ground is permanently waterlogged. A spongy carpet of sphagnum moss grows on top of dead moss slowly forming peat. It is a delicate environment with the moss like a skin on the bog. Both areas are protected as Sites of Special Scientific Interest [SSSIs]. Where it is less wet, there are heather and grass moors. In winter, patches of old heather are burnt off to encourage new growth and grazing for animals. Some areas have too many animals and the environment is suffering.

The national park authority has monitored Dartmoor's environments and habitats for more than 20 years. It has bought land around Haytor to carry out research into the best ways to manage the moorland. One of the aims of the 'Moor Care Project', partly paid for by the European Union, is to educate visitors and local people to look after Dartmoor.

TASKS

1 Make a full-page copy of the map of Dartmoor (Figure 4.21). Annotate it with the information from this section.

2 Should the army train on Dartmoor? In class, stage a 'public inquiry' putting forward arguments for and against. Finish with a vote.

3 Some areas are overgrazed, but farmers must make a living. What are the options for the national park authority which wants to conserve the grass moors?

4 Look at Figure 4.22. Do you think the reservoir spoils or enhances the landscape? Write down your reasons.

You can find out more detailed information on the Council for National Parks' website:
http://www.dartmoor-npa.gov.uk/

5 Using the information in this book together with the website, produce a desktop published leaflet outlining the land use issues facing Dartmoor National Park.

WATER POLLUTION

Sources of water pollution

Water in rivers, lakes or the sea can be polluted by:

▶ **Sewage** In MEDCs, domestic and industrial drains usually carry sewage to treatment works which discharge the treated sewage into rivers, lakes and seas. Some old treatment works are inefficient and leaks of raw sewage happen, for example when drains overflow in a flood.

▶ **Manufacturing industry** Many factories are located alongside water and use the water for cooling purposes, with the effect that warm water is returned to the river, lake or sea. The factories may deliberately or accidentally discharge unwanted by-products of their industrial processes into the water.

▶ **Agriculture** Farmers apply nitrate fertilisers, herbicides (weed-killers) and pesticides (pest control) to their crops. Excess chemicals may be washed into rivers and lakes – and on into the sea. Cattle slurry and silage (fermented grass for fodder) are stored on farms and may leak into nearby bodies of water.

▶ **Energy** Many power stations use water for cooling and then discharge warm water. Power stations that burn coal or oil also produce acid rain, which affects rivers and lakes (see pages 202–205). Spills of crude oil occur in estuaries and seas as energy is moved around the world in tankers.

The effects of sewage pollution

▶ Bacteria break down sewage into ammonia but use up oxygen from the water in the process. The lack of oxygen means a smaller range of creatures can live in the water.

▶ Other bacteria break down the ammonia into nitrates, which leads to **eutrophication**.

▶ As the plant life encouraged by eutrophication grows, dies and is broken down by bacteria, more oxygen is used up.

The effects of nitrate pollution

▶ Nitrates are plant food, and too much of them will encourage the growth of green plants which use up oxygen and block out light, thereby reducing the range of creatures which can live in the water.

▶ The EU is concerned about a threat to health from high nitrate levels in drinking water and has set a limit of 50 mg/litre. This can only be met by expensive treatment of water supplies or by reducing the use of nitrate fertilisers by farmers. UK consumption of nitrate fertilisers increased from 20 000 to 1 600 000 tonnes per year in the 40 years after 1948.

TASKS

1 Read about the pollution incidents below. Make a larger copy of the table below.

Write each case study's name in the appropriate place. Some may spread across two rows or columns:

Amoco Cadiz 1978 This oil tanker ran aground off the Brittany coast, France, and leaked crude oil.

Sandoz 1986 A fire at this chemical factory in Basle, Switzerland, caused chemicals to be washed into the River Rhine when firefighters fought the blaze with hoses; the pollution flowed downstream and into the North Sea.

Bulmer's 1994 The Hereford cider manufacturers admitted polluting the River Wye with an accidental leak of liquid glucose which caused a loss of salmon stock.

River Stour, East Anglia Nitrate levels rose steadily from 1940 to 1980 towards the EU limit.

Southern North Sea 1989 A toxic algae plague was reported off Britain's coast; it was blamed on eutrophication due to pollution from farms and sewage works and it seriously damaged fish stocks.

Wheal Jane tin mine, Cornwall – overflow water from the disused mine carried heavy metals into local rivers.

Source	Lake	River	Sea
Sewage			
Agriculture			
Industry			
Energy			

SPANISH SLUDGE SPILL, APRIL 1998

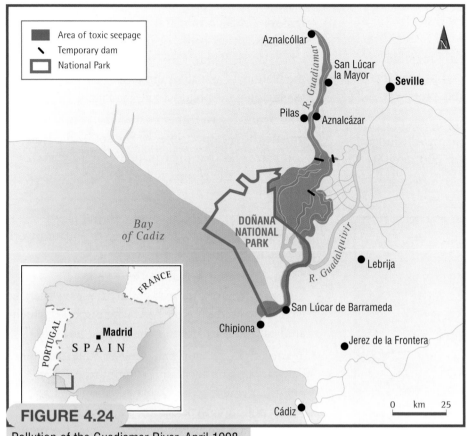

■	Area of toxic seepage
➚	Temporary dam
▭	National Park

FIGURE 4.24

Pollution of the Guadiamar River, April 1998

Causes of the pollution

A wave of toxic mining waste poured into the headwaters of the Guadiamar River, Spain, when the dam holding back a reservoir of mining waste from the Los Frailes mines burst on 25 April 1998 (see Figure 4.25). The wave consisted of an estimated 5 million cubic metres of acidic sludge containing poisonous lead, arsenic, zinc and mercury. Boliden, a Canadian/Swedish company, owners of the Los Frailes zinc and lead mines at Aznalcollar, claimed that the dam gave way because of a 'seismic shift' but other people said that the authorities had been warned that the dam was unstable and likely to burst.

FIGURE 4.25

The broken dam which released the toxic mining waste

Short-term effects of the pollution

‣ 400 ha of farmland were covered by toxic sludge (see Figure 4.24); the damage to crops growing in olive groves, orchards and vegetable plots was estimated at £6 million.

‣ The Guadiamar river was blackened with sludge for about 40 km and to a width of 200 m either side of its banks.

FIGURE 4.26

Fish die in the toxic sludge

‣ Large numbers of fish, crustacea and other invertebrates were killed and there was a fear of a widespread knock-on effect through the food chain, causing deaths of birds and mammals.

‣ 400 families had their water supplies polluted.

‣ Homes alongside the river were flooded with polluted water.

‣ People living in the affected area reported itching of their eyes, rashes and difficulty in breathing.

‣ The salt marshes of the Donana nature reserve lie downstream of the sludge spill and were threatened by pollution and poisoned food being washed in. Donana is Europe's largest wild bird sanctuary, with a quarter of a million birds nesting there.

Longer-term effects

‣ The authorities were criticised for what was seen as a slow response to the disaster.

‣ The mining company mounted an operation to remove 7 million tonnes of topsoil from the Guadiamar river banks before the autumn rains which threatened to carry the pollution further into the National Park; all of the vegetation was removed, leaving only the polluted soil (see Figure 4.27).

‣ The Spanish government asked the courts to pursue those responsible for the spill. The Boliden company said that their insurers would cover any costs which the company had to pay as a result of the spill.

‣ The mine engineers continued to pump waste into the River Guadiamar so that they could work on repairing the broken dam.

‣ There were concerns about long-term effects on health. Groundwater was polluted. Heavy metals were deposited on river beds where species which are near the start of food chains feed and breed.

‣ Teams of citizens and park rangers built makeshift dams in a partly successful attempt to prevent pollution entering the Donana National Park. Much pollution bypassed the Park via the Guadalquivir River (see Figure 4.24); at the mouth of this river nearly 1 000 small boats collect fish and seafood including shrimps and eels.

FIGURE 4.27

Cleaning up the toxic sludge deposited in the Guadiamar valley

FIGURE 4.28

The area before the spill, 8.3.98

FIGURE 4.29

The area after the spill, 25.4.98

▶ Hundreds of workers gathered 20 tonnes of dead fish in and around the National Park in an attempt to to prevent them being eaten by birds, and Park officers scared birds away with guns.

▶ The delicate wetlands of the Donana National Park faced a water shortage because they could no longer take water from the polluted Guadiamar River.

▶ Consumer groups issued warnings about the risks of buying poisoned local fruit, vegetables, fish or seafood from illegal sellers in the area.

▶ Farmers' organisations said that the land poisoned by the toxic flood might have to be left fallow for as much as 25 years.

▶ Government scientists suggested treating the acid water with lime and introducing special plants in the region to absorb heavy metals.

▶ A popular annual pilgrimage – which involves wading across the River Guadiamar – faced disruption. The Spanish Army was brought in to provide a temporary bridge across the polluted river.

TASKS

1 a Trace the frame of the satellite image for 25 April (Figure 4.29). Then, using Figure 4.24 to guide you, trace the following features:
- the coastline
- the Guadalquivir River
- the Guadiamar River.

 b Use colour shading to mark the following details onto your tracing:
- the reservoir from which the toxic sludge originated
- the area affected by toxic seepage
- the area of the Donana National Park
- the urban area of Seville.

(To find out more, access:
http://www.dfd.dlr.de.lapp/iom/1998-07/index.html.en)

There are a number of environmental groups which are currently engaged in campaigns to reduce levels of water pollution around the planet. Their viewpoints about these issues are accessible on their websites, for example:

WWF Global Network http://www.panda.org/
Greenpeace International http://www.greenpeace.org/
Friends of the Earth http://www.foe.org.uk/

Using information obtained from this textbook together with websites, produce a PowerPoint presentation for the rest of the class, which outlines the causes and consequences of water pollution.

PEMBROKESHIRE COAST OIL SPILL, WALES

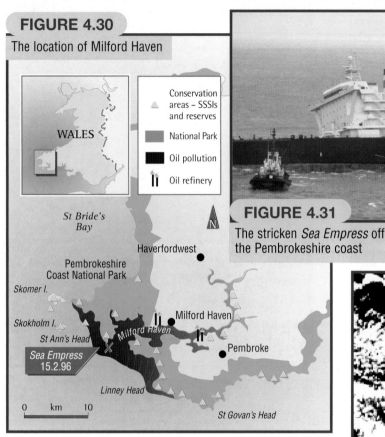

FIGURE 4.30

The location of Milford Haven

WALES

Conservation areas – SSSIs and reserves

National Park

Oil pollution

Oil refinery

St Bride's Bay

Haverfordwest

Pembrokeshire Coast National Park

Skomer I.

Skokholm I.

St Ann's Head

Milford Haven

Milford Haven

Pembroke

Sea Empress 15.2.96

Linney Head

St Govan's Head

0   km   10

FIGURE 4.31

The stricken *Sea Empress* off the Pembrokeshire coast

▶ the coastline within Milford Haven was heavily oiled; much of the oil moved south and then east, affecting the coastline as far as Pendine Sands.

▶ some oil reached Skomer Island;

▶ Lundy Island in the Bristol Channel received light oiling;

▶ tar balls were blown across the sea as far as the Irish coast and the north Devon coast.

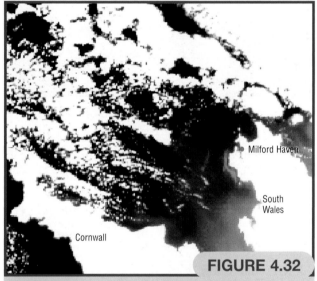

Milford Haven

South Wales

Cornwall

FIGURE 4.32

Satellite image of south-west Wales taken at 08:08 on 21 February 1996, at the height of the pollution incident

The *Sea Empress* oil spill

On the evening of Thursday, 15 February 1996, the 147 000 tonne supertanker *Sea Empress* approached Milford Haven with a cargo of light crude oil from the North Sea. The ship attempted to enter the port at low tide on its way to the Texaco refinery and ran aground off St Anne's Head just after 20:00. Two hours later, the *Sea Empress* was afloat again, having lost about 6000 tonnes of oil through the damaged single-skin hull. Conflict and confusion among those responsible for salvage meant that it was over a week before the vessel was being unloaded at a jetty in the Haven; during this time the vessel had broken away from tugs, run aground several more times, been battered by a storm and leaked more than 70 000 tonnes of oil into the marine environment – roughly half of its original cargo.

The oil spread from the leaking *Sea Empress* was carried by winds, currents and tides.

Responses to the oil spill

A significant proportion of the oil at sea was removed naturally by evaporation, some was recovered mechanically by skimming but the main method used was to spray chemical dispersant from the air; this began only 17 hours after the *Sea Empress* started leaking oil, using the UK's fleet of seven specially adapted aircraft. Between 17 and 25 February, 445 tonnes of chemical dispersant were used to break up the oil and remove it from the surface, thereby reducing the hazard to seals and seabirds. Some observers were worried about the poisonous effect of the dispersant itself.

Popular tourist beaches such as at Tenby have great economic importance for the tourist industry, so 'aggressive' cleaning techniques were applied. These may have been harsh on the environment. Pressure-washers directed oil towards trenches dug in the beach, from which pumps
lifted the oil into skips. Most of the main bathing beaches were re-opened for Easter.

FIGURE 4.33 Beach cleaning in progress

On less intensively used stretches of coast the natural processes of dilution, dispersal and degradation were boosted. The experience of the *Exon Valdez* tanker spill in Alaska in 1989 indicated that washing beaches with high-pressure hoses can drive the oil deeper into the beach sediments, where it might remain for years. Natural wave action gradually washes oil from pebbles to be broken down by naturally occurring bacteria. An experiment was set up at Bulwell Bay to investigate the effectiveness of bioremediation by burying 'socks' full of nutrients in the beach to accelerate bacterial action. The experiment appeared to be a success: the rate of decay was doubled. The technique is not appropriate for amenity beaches, which need a faster clean-up rate than that which bioremediation can achieve.

Effects of the oil spill

Wildlife

Marine mammals such as seals and dolphins appeared to be little affected by the spill, but birds were hit hard with 20 000 dead. Sea birds suffer because they fail to identify the oil as a hazard and get their plumage coated; the oil reduces their insulation and waterproofing and in their attempts

to clean themselves by preening, they swallow oil which poisons them. The winter population of scoters in Carmarthen Bay fell from 14 500 to 4000.

It was disastrous for shellfish, which are near the bottom of the marine food chain and feed by filtering sea water through their bodies. Limpet populations suffered 50% losses, with 90% being recorded at one site. One effect was that exceptional amounts of algae covered some rocky shores in July because the herbivorous limpets were not there to eat it. Hundreds of bivalves (e.g. cockles and razor shells), urchins and starfish were washed up on beaches. Plant communities of saltmarshes, shingle, sand dunes and cliffs were damaged.

Fishing

Immediately after the oil spill, catching fish and shellfish was banned. Commercial fisheries here had earned over £3 million a year and employed over 1000 people.

Tourism

Many local people are employed in catering, accommodation and leisure activities; sport fishing alone was estimated to be worth over £1 million a year. As news of the oil spill spread, requests for holiday brochures decreased, as did bookings. Losses to tourism and fisheries are estimated at between £13 million and £50 million.

Farming

Onshore winds blew an oily film inland, which settled on some farmland. Experts decided there were no food safety implications.

TASKS

1 Compile a case study revision map of the Pembrokeshire oil spill. You should draw an outline map of the coastline with plenty of space around it to make notes which should summarise the information on pages 200 and 201.

2 Arrange the following words to create three food-chain webs found in a coastal environment: mudflats, limpets, hawks, insects, marsh grass, frogs, algae, bivalves, rocks, saltmarsh, wading birds.

3 Put yourself in the position of each of the following Pembrokeshire people and write down no more than four short points to summarise the likely views of each person on the *Sea Empress* oil spill:

a Tenby hotel-owner	a wildlife trust volunteer
a commercial fisherman	a field study centre warden
a coastguard	an oil refinery worker

THE CAUSES OF ACID RAIN

The pH scale measures acidity (see Figure 4.34); the scale is logarithmic, which means that pH 4 is 10 times more acidic than pH 5 and 100 times more acidic than pH 6. 'Normal' rain is slightly acid because of naturally occurring gases which dissolve in water in the atmosphere; these gases come from sources such as plant respiration, volcanic eruptions and forest fires. **Acid rain** is the name given to any precipitation with a pH value below 5.6; the additional acidity comes mainly from two widespread human activities:

▶ power stations burning fossil fuels which have a high sulphur content. These add the gas **sulphur dioxide** to the atmosphere;

▶ moving around in vehicles which emit **nitrogen oxides** from their exhausts. These gases are also emitted when fuels are burned in power stations.

The polluting gases react with oxygen in the atmosphere to produce dilute forms of **sulphuric acid** and **nitric acid**. These acids are taken into solution by cloud and rain droplets and then move through the hydrological cycle.

FIGURE 4.35

Emissions from Ironbridge power station, Shropshire

The acids may travel 1000–2000 km in the atmosphere in 3–5 days, carried by the **prevailing winds** (winds which regularly blow from a similar direction). Infrequent summer rains may have accumulated more acid than winter rains because of a dilution effect in winter. When acid rain falls, it lands on vegetation, soaks in to the soil and enters rivers and lakes.

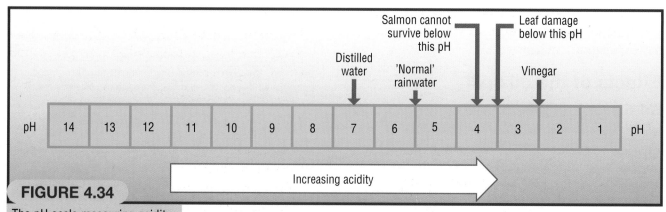

FIGURE 4.34

The pH scale measuring acidity

TASKS

1 Study Figure 4.36, which shows data on European pollution gathered more than 20 years after the 1972 United Nations conference in Stockholm which alerted the world community to the problem of acid rain.
   a In rank order, list the countries which were net exporters of sulphur (emission greater than deposition).
   b In rank order, list the countries which were net receivers of sulphur (deposition greater than emission).
   c Suggest reasons to explain the pattern in your ranked lists. Use atlas maps which show the physical and economic geography of Europe to help.
   d Use the same atlas maps to attempt to explain the variations in sulphur emissions per person.
   e Which countries do you think were pushing hardest for action against acid rain – and why?
   f Why do you think it is difficult for one country to blame another for being the source of acid rain?

Country	Total emission	Total deposition	Emission (tonnes/1000 people/year)
Belgium	152	62	15
Czech Republic	710	260	69
Denmark	78	45	15
Finland	60	106	12
France	568	362	10
Germany	1 948	803	24
Italy	1 126	299	20
Netherlands	84	71	5
Norway	18	94	4
Poland	1 362	822	35
Sweden	50	161	6
United Kingdom	1 597	430	27

Figure 4.36 Sulphur emissions and depositions in Europe (2 tonnes of sulphur dioxide is equivalent to 1 tonne of sulphur) – figures are in 1 000 tonnes per year in 1994

THE EFFECTS OF ACID RAIN

▶ Rivers and lakes become more acidic. Animal life decreases in lakes and streams as acidity increases. Out of 90 000 Swedish lakes, 18 000 have been seriously damaged by acidification and 4 000 have no life in them. Some lakes and streams in southern Scotland, the English Lake District and in mid-Wales have been affected too. Some upland lakes and streams appear to have 'surges' of acidity as substantial quantities of acid stored in frozen snow are released on melting in spring.

▶ Acid rain infiltrates the soil surface. As water moves through the soil, chemical reactions release toxic metals – particularly aluminium – into solution. The resulting **acid-toxic stress** may damage soil bacteria, tree roots and – when the water returns to streams and lakes – fish. The acidification of groundwater in Sweden has been linked to the corrosion of water pipes, pots and pans, foul-tasting water, children's diarrhoea, and hair turning green after washing! In some badly affected parts of the Czech Republic, it is estimated that sulphur dioxide pollution has reduced life expectancy by 10 years compared with the national average. In the Yorkshire Pennines, an additional two stages of filtration had to be added to the water treatment process for domestic supplies because of increased metal content.

▶ Acid rain damages trees, but they may also be affected by fungus, drought and disease and it is therefore difficult to single out acid rain as the cause of their ill-health. A survey of forests in Germany revealed that 2 500 000 ha of forest were damaged (1 in 12 trees were affected) but scientists were unable to reproduce the symptoms in experiments. It is believed that trees may suffer from stress. This results from a combination of acid water falling onto foliage and the effect which acid soil water has on nutrient uptake by tree roots.

▶ Food crops may be affected by acid rain. More than 1 000 ha of rice were killed by acid rain near Chongqing in China.

▶ Acid rain damages stonework in the built environment; in most urban areas it is possible to see stonework that has been blackened and eroded by the chemical action of acid rain. It has been estimated that the Acropolis – an ancient monument in Athens – has deteriorated as much in the last 20 years as in the previous 2000 years.

These effects have serious financial consequences for fishing, tourism, farming, forestry, water supply and building maintenance in the countries affected by acid rain. The annual cost of damage from acid deposition in Europe is estimated between $0.5 and $3.5 billion.

Another important effect of Europeans' concern about acid rain is that coal is now seen as a 'dirty' fuel. Coal and oil are losing out to gas as fuel for generating electricity. In the UK there has been a 'dash for gas' since the privatisation of electricity generation; during the 1990s gas has risen towards meeting a third of the UK's electricity demand because the generating companies see gas-fired power stations as more profitable (page 161). Environmental arguments can be made for gas-fired power stations because they emit less pollution than coal-fired power stations: half the carbon dioxide, a tenth of the nitrous oxide and negligible amounts of sulphur dioxide. Britain's coal-mining industry was devastated as the main electricity generating companies cut their coal orders by more than half in the five years to 1997. This brought enormous economic and social costs to Britain's coal-mining communities (page 160).

TACKLING ACID RAIN
International co-operation

It is impossible to blame acid rain from one particular country for causing environmental damage in another because of the complicated – and disputed – causes of the damage. It is widely recognised that the way forward is through international co-operation, and this is happening in Europe.

▶ The first European international agreement was signed by 31 countries in 1979 and proposed collaboration in research, pooling of information and effective reductions in sulphur dioxide emissions;

▶ A later UN proposal was signed and agreed by fewer countries (not including the UK), some of which went on to achieve a 30% reduction in sulphur dioxide emissions;

▶ In the 1980s, UK electricity generators and coal producers started a joint study with Norway and Sweden into the processes which cause acid rain and its associated problems.

Many of the proposed reductions fall short of what environmentalists want to protect the environment. There has been a reduction in sulphur dioxide emissions across Europe but nitrogen oxide and carbon dioxide emissions have continued to grow, although at a slower rate than anticipated. Some people believe that the improvements may be less to do with anti-pollution measures and more to do with:

▶ a reduced demand for energy because of economic recession;

▶ a switch from coal and oil to gas and nuclear fuels for economic and political reasons.

FIGURE 4.37

Tree damage attributed to acid rain in the Erzgebirge Mountains, Czech Republic

Reducing emissions from coal-fired power stations

The following measures have been taken in the UK:

▶ Flue Gas Desulphurisation (FGD) equipment – sometimes known as 'scrubbers' – can remove up to 95% of sulphur dioxide emissions through a chemical reaction with crushed limestone; the equipment is expensive and has not been fitted to many UK coal-burning plants.

▶ Low-sulphur coal may be blended with the traditional high-sulphur coal before burning. Crushing and washing coal has been found to reduce sulphur dioxide emissions by up to 15%.

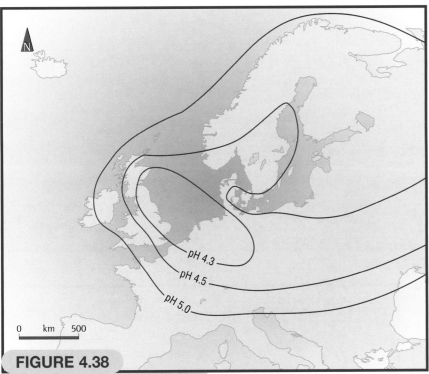

FIGURE 4.38

Average acidity of rain across Europe

▶ Low temperature burners cut emissions of nitrogen oxides. These have been installed in some coal-burning power stations in the hope of reducing the nation's nitrogen oxide emissions by 30% over 10 years.

All new coal-fired power stations are to be fitted with FGD equipment and low temperature burners. It is expensive to fit emission-reducing equipment to existing power stations. To reduce UK emissions by 60% of the 1980 level by 2003 is estimated at between £1 billion and £2 billion.

Other solutions to the problems of acid rain

▶ Lime can be added to lakes affected by acidification. This has been done in Sweden for many years. However, it has damaging effects on the lakeside animals and plants.

▶ Alternatives to coal and oil are being used: in the UK, the electricity suppliers have to buy some of their energy from other sources. A main alternative is nuclear energy (with its associated problem of waste disposal) but small contributions are made by wind power, hydro-electricity, and landfill gas plants.

▶ Most filling stations in Europe now sell low-sulphur diesel fuels for use in cars and commercial vehicles.

▶ Energy conservation is becoming increasingly important. The EU has an energy conservation programme called SAVE (Specific Actions for Vigorous Energy Efficiency) which demands higher energy efficiency in new vehicles and homes. Energy conservation may be achieved in homes through steps such as better insulation of roofs, walls, windows and doors, more efficient central heating boilers and 'passive solar heating' by careful positioning of windows in relation to the sun.

▶ Buildings are being scrubbed clean to remove the discolouration created by more than a century of acid rain and some are being treated with anti-corrosion films.

TASKS

1 Use Figure 4.38 as the basis for an argument on behalf of Norway and Sweden against coal-burning power stations in the UK. Present your argument in the form of an annotated sketch map which shows acid rain, prevailing winds and UK coal-fired power stations. You will have to look at an atlas to find some of this information.

2 Put forward the argument which the UK might have used in 1979 to counter the argument put forward by Norway and Sweden.

GLOBAL WARMING

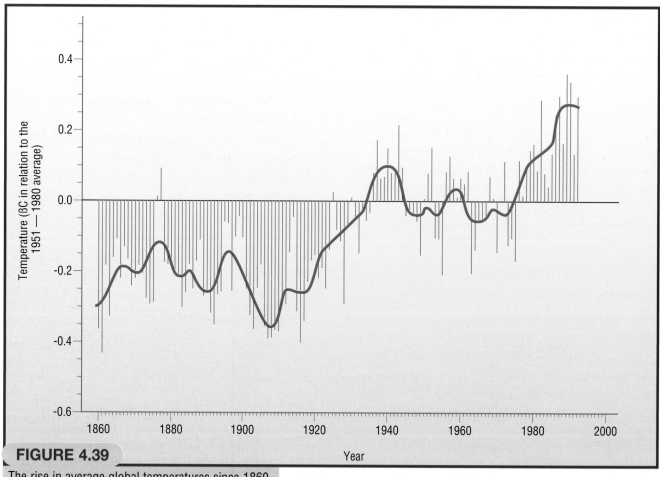

FIGURE 4.39

The rise in average global temperatures since 1860

Climate change

Average global temperatures have been fluctuating for millions of years. They are affected by variations in incoming solar radiation, movement of the earth's crustal plates, changes in ocean currents and atmospheric circulation and changes in the gas content of the atmosphere. We know about temperatures millions of years ago because of evidence provided in sea-floor sediments and ice-caps. Figure 4.41 shows a definite rise in temperature during the twentieth century but there is debate about how much of this rise is due to our consumption of resources and energy and how much is due to a 'natural' fluctuation.

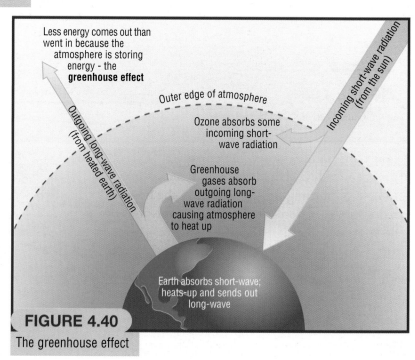

FIGURE 4.40

The greenhouse effect

The greenhouse effect and global warming

The earth's atmosphere is mostly made up of a mixture of nitrogen and oxygen, with much smaller amounts of heat-retaining gases referred to as **greenhouse gases** (see Figure 4.43). Figure 4.40 shows how the **greenhouse effect** operates.

‣ Energy travels through space from the sun as short-wave radiation which passes through the most of the gases in the earth's atmosphere without being absorbed (except that some harmful ultra-violet radiation is absorbed by a gas called ozone);

‣ The short-wave radiation heats the earth's land and sea surfaces which then re-radiate the energy back towards space as long-wave radiation;

‣ The greenhouse gases absorb large amounts of this long-wave radiation, which causes the earth's atmosphere to heat up.

The theory of global warming is that human activities are increasing the amount of greenhouse gases in the atmosphere and this is causing more and more long-wave radiation to be trapped. This results in a further rise in average global temperatures.

THE POTENTIAL CONSEQUENCES OF GLOBAL WARMING

‣ Sea levels may rise because ice caps and glaciers melt and add more water to the world's oceans. Also water expands as it gets warmer. Estimates of the rise in sea level vary, but low-lying areas are threatened by a relatively small rise of 50 cm over the next 100 years which would flood 12% of Bangladesh, deprive Egypt of 20% of its arable land; and threaten many small islands.

‣ Effects on climate are complex and subject to dispute among experts, especially when El Niño is considered. El Niño is an unusually warm current in the Pacific Ocean and appears to increase climatic hazards. Global warming may favour El Niño, bringing droughts to Africa, south Asia and Oceania and more powerful hurricanes to the east Pacific and the Caribbean. One bizarre effect of global warming could be that the UK could eventually become much colder because changes in the ocean currents might divert the North Atlantic Drift, which warms the British Isles. Another possible effect is that climatic belts would move polewards by up to 250 km. A UK government report concluded that southern England would inherit the warm climate of the Loire Valley (France) in 25 years.

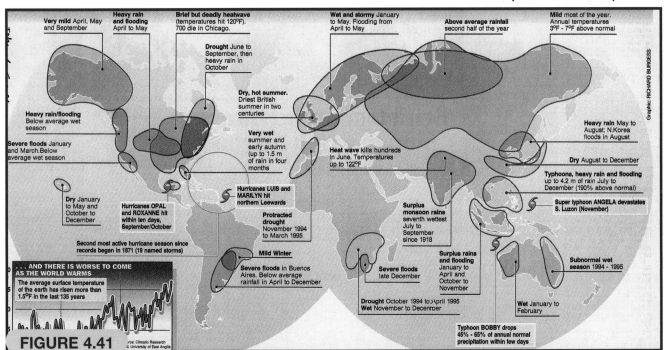

FIGURE 4.41

The world's climatic extremes in 1995, believed by some experts to be evidence of the effects of global warming

▸ Agriculture would be affected by climatic change. It might be possible to grow vines in southern England on a much larger scale than at present. Global food production may remain much the same, but densely populated, poor areas of the world might suffer as much as a 20% drop in harvests, thereby unbalancing food supplies and demands even further than the present situation.

▸ Natural vegetation and animals might not adapt or migrate quickly enough to keep up with the changing climate. One-third of the world's forests could be at risk, some from forest fires which might add more carbon dioxide to the atmosphere, thereby amplifying the greenhouse effect. The natural habitats of some animals could be lost completely, resulting in a loss of species diversity.

▸ Many tourist resorts depend on narrow climatic or coastal margins: if average temperatures rise, snowfall may decrease in the Alps, devastating the winter ski-ing industry; holiday developments in LEDCs rely on attractions such as wide beaches and coral reefs which may disappear with rising sea levels and changing water temperatures.

▸ Health problems are often associated with climate: tropical diseases such as malaria are expected to alter their scale and geographical distribution, moving into temperate latitudes.

▸ The financial costs of global warming are enormous, from insurance claims for hurricane damage to the cost of irrigating areas which have suffered a loss of rainfall.

TASKS

1 Many people believe that increased greenhouse gases are causing global warming. Using Figure 4.44 as a starting point, explain why the concentration of greenhouse gases has increased during the twentieth century.

Greenhouse gas	Source	Estimated % contribution to global warming
Carbon dioxide	Burning fossil fuels to provide energy and burning vegetation, e.g. deforestation.	55
Methane	Produced by bacterial decay of organic material; released by swamps, paddy fields, the digestion of grazing animals and by mining (having been trapped in coal for millions of years).	15
Chloro-flouro-carbons (CFCs)	Synthetic chemicals used to propel aerosol sprays, manufacture plastic foam packaging and as a refrigerant.	24
Nitrous oxide	Released when farmers use artificial fertilisers and when fossil fuels are burned in power stations and vehicles.	6

Figure 4.43 Greenhouse gases

2 Use the information on these pages as a starting point to list different ways in which each of the following outcomes might arise from global warming:

flooding; damage to natural ecosystems; additional costs to governments, organisations and individuals.

HOW TO DEAL WITH THE EFFECTS OF GLOBAL WARMING

International co-operation

Global warming is an issue which demands international co-operation and agreement. The Intergovernmental Panel on Climate Change (IPCC) was set up in 1988 by the United Nations and the World Meteorological Organisation.

By 1995, the IPCC had decided that there was enough reliable evidence to conclude that pollution is at least partly to blame for the earth heating up. It was the subject of the UN conference on greenhouse gas emissions and climate change at the end of 1997 in the Japanese city of Kyoto (see Figure 4.44).

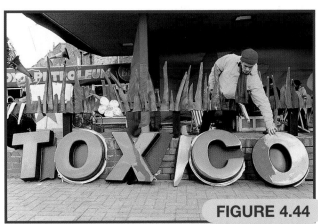

FIGURE 4.44

Protesters occupy a derelict petrol station in London during the Kyoto summit to protest about the world's high consumption of fossil fuels and consequent emission of greenhouse gases.

The targets at Kyoto were meant to be legally binding but it has always been difficult to persuade the USA – responsible for 25% of world emissions – to make meaningful cutbacks. There is the ethical dilemma that global warming has been caused by the industrialisation of MEDCs but if LEDCs are going to raise their own living standards, they will need to industrialise and increase their emissions. It is argued that the solution is to achieve a convergence towards equal emissions per person by 2050 – a big reduction by MEDCs and a reduced rate of increase in LEDCs.

Possible ways of reducing emissions

- Promoting greater energy conservation and efficiency, for instance a series of TV adverts in the UK promoting individual energy-saving actions and a road tax reduction for cars with engines under 1100 cc, which consume less fuel and consequently emit less carbon dioxide.
- Phasing out CFCs: most aerosol sprays now use an alternative propellant and many local authorities subsidise a scheme which carefully disposes of the CFCs contained in discarded fridges.
- Increasing the use of renewable energy: in 1995 the UK generated only 2% of its electricity from renewable sources, compared with the EU average of 14%. The government says it wants to reach 10% by 2010. Average costs for wind, hydro, landfill gas and waste-burning fell dramatically during the 1990s. Some electricity suppliers allow environment-conscious customers to pay about 10% more on their bills to fund renewable energy projects.
- Halting deforestation and encouraging tree-planting because trees turn carbon dioxide into oxygen.
- Changing agricultural practices to reduce methane and nitrous oxide emissions.

TASKS

1 Draw up a list of the actions which individuals and organisations could take to reduce emissions of greenhouse gases directly or indirectly.

2 Suggest reasons why it has proved difficult to achieve reductions in emissions of greenhouse gases in MEDCs.

There are a vast number of Internet sites which convey different viewpoints about global warming. Two examples are provided below:

Global Warming Information Page:
http://www.globalwarming.org/

Friends of the Earth Climate Change Campaign:
http://www.foe.co.uk/climatechange/index.html

3 Use information from these sites to produce your own desk top published flyer about global warming. Outline the causes, consequences of the problem as well as suggested solutions.

COURSEWORK FOR THE ENVIRONMENTAL EFFECTS OF PEOPLE

Opportunities for coursework based upon this part of the syllabus are not equally spread. There may be no local opportunities. While rivers of a suitable size for a study of pollution and its effects are more widely available than quarries, open cast mines and national parks, rivers have been so cleaned up in recent years that fieldwork may yield no significant results. Where there are plans to start up open cast mining or increase quarrying potentially there is a local issue worthy of investigation. The National Parks of England and Wales are so popular with visitors that in all of them there are conflicts of interest or evidence of tourist over-use, especially in honeypot locations.

River pollution

Sites will need to be chosen carefully and researched for their suitability. There may be opportunities if readings are taken above and below an outflow pipe, or in slower flowing, more stagnant stretches of the stream. Make a booking sheet similar to the one shown in Figure 4.45. As always when recording in the field, the booking sheet needs to start with spaces for essential details about the site. After that you can take certain measurements, such as width, depth and velocity previously referred to on pages 54–55. Test acidity by using pH paper and assess clarity by putting a white yoghurt carton on the end of a pole and lowering it into the water. Record the depth at which the carton is no longer visible.

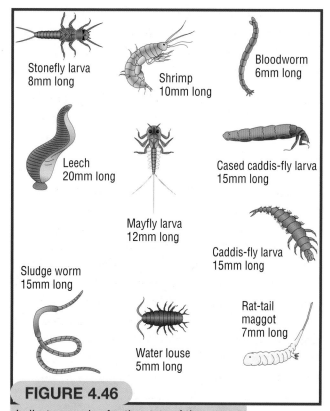

FIGURE 4.46

Indicator species for the state of the water

If you are interested in biology, the study can be extended to survey the water life in spring and summer. Figure 4.46 gives some indicator species to look for. Try to count the number of different species at each site. Expect to find about 20 in non-polluted water.

River Pollution Booking Sheet			Overall score: 0-3 very clean 4-9 clean 10-15 fairly clean 16-21 doubtful 22+ badly polluted		
River, stream		Site Number		Grid reference	
Date		Weather		Depth of river (cm)	
Width (m)		Ave. surface velocity (m/sec)		Visibility using yoghurt carton (cm)	
Water temperature (°C)		pH			
Visual Survey (✓ tick)	**0 points**	**1 point**	**2 points**	**3 points**	**4 points**
Presence of suspended solids, e.g. sewage	Very clear	Clear	Fairly clear	Slightly murky	Murky
Colour	Very clear	Clear	Slightly brown	Dark brown	Black
Stones	Clean and bare	Clean	Lightly covered in brown fluffy matter	Coated with brown fluff	Coated with brown and grey deposits
Water weed	None	A little in shallows	Lots in shallows	Abundant	Chocked
Grey algae (sewage fungus)	None	None	A little	Present in patches	Plentiful
Scum/froth/oil	None	Odd bubbles	Noticeable foam islands	Large quantities	Covers whole river
Dumped rubbish	None	A few small items	A few large items	Large and small items	Many large, different items

FIGURE 4.45   Recording sheet for river pollution

Mining and quarrying

Mining and quarrying are controversial land uses in rural areas where they spoil the natural beauty and increase noise, air and water pollution and traffic levels. Yet they may be significant sources of employment. It is important to seek the views of more than one group of people, for example both locals and visitors, to gain a balanced picture of benefits and disadvantages.

Figure 4.47 shows student ideas for investigating the impact of an open cast mine, which can act as a starting point for a quarry study as well.

National Parks

One approach to coursework on national parks is to concentrate on one honeypot location, whether a settlement or a scenically attractive natural site. Begin by explaining the physical and human factors which have led to it becoming a honeypot. Undertake fieldwork to discover the evidence that shows whether the area is being over-visited or over-used. The type of evidence will depend on the time of the visit. On a sunny day in summer visitor pressure can be measured by people and traffic counts and by counts of numbers of cars parked in both official and unofficial places. There will be plenty of customers for your questionnaires,

TITLE OF THE STUDY

'What is the impact of open-cast mining on people and the environment?'

Observation
* Mapping the site and its workings including heaps of overburden and waste.
* Environmental survey of the site and surrounding area.
* Environmental survey of areas unaffected for comparison.

Measurement
* Vehicle counts during working hours.
* Noise levels recorded on a sound meter (around the site and further away).

Interview / Questionnaires
* Arrange an interview with the site manager to hear the company's view and gain details such as the number of workers and home locations, where the raw materials from the mine go to, the number of lorries and routes, and pollution precautions.
* Questionnaires to local residents and farmers about their opinions of the site. Ask people in different locations at different distances from the site.
* Interview owners of local shops and services - has the mine brought any extra income?

Secondary sources
* Geological maps
* Newspaper articles
* Open-cast mine applications and environmental reports

FIGURE 4.47

A student plan for data collection

which can be modified versions of the one given on page 169. Out of season it is easier to concentrate on signs of over-use, especially footpath erosion. One footpath to a popular site can be followed. Its width and depth can be measured, and its surface features noted, and compared with physical features. At the same time methods of footpath management can be studied.

FIGURE 4.48

Measuring footpath erosion in winter. Quadrats are useful for estimating the percentage of bare ground / vegetation cover from one side of the path to the other.

EXAMINATION TECHNIQUE – UNIT 4

In Unit 4 an emphasis is put on the relationship between people and the environment. Examination papers reflect this link, with questions that investigate environmental issues at all scales, from local to global.

This relationship between people and the environment also stresses the different viewpoints or attitudes of people towards the environment, and how these can influence the way that they interact with it.

In order to answer questions on environmental topics successfully you need to have not only knowledge of geographical issues and case studies, but also an understanding of how people may have different ideas about how the natural environment should be used, and how this use should be managed in order to protect the environment.

The first question tests your understanding of a global environmental issue – acid rain. The same resource is used in both the Foundation and Higher Tier papers.

The diagram can be used to answer the following questions.

Foundation Tier question

1 (a) Use the ideas in Fig. 1 to explain why acid rain occurs. [3]

Higher Tier question

2 (a) Use the ideas in Fig. 1 to describe the effects of acid rain and explain why it occurs. [6]

There is a clear difference between the two questions. Whereas the Foundation Tier question instructs the candidate to *explain*, the Higher Tier question has two command words, *describe* as well as *explain*. This means that the candidate has to do two tasks in order to score maximum marks. If you only describe the effects or explain why acid rain occurs you will be limited to scoring only up to four marks. The remaining two marks are what are called 'reserve' marks, which means they are reserved for completing the other task.

The second part of this question is also similar on both Foundation and Higher Tier papers.

Foundation Tier question

3 (a) Explain why countries need to work together to reduce acid rain. [3]

Higher Tier question

4 (a) What can be done to reduce acid rain ? Explain why countries need to work together to achieve this. [6]

The Higher Tier question is again more difficult because candidates are first required to answer the general question about environmental protection and then explain the need for international cooperation. Foundation Tier candidates need only deal with international co-operation.

From this question you can see that when revising a geographical **concept** such as acid rain it is possible to break the topic down into three simple questions.

▶ How is the problem caused?

▶ How does it affect people and/or the environment?

▶ What can be done to reduce or manage the problem?

Before answering an examination question, read it carefully. Often you will not have to deal with all three of these questions.

The following example shows how such a revision plan would work in order to answer a question about global warming (or as it is often called the greenhouse effect). Like acid rain, global warming is an international or global issue. This question is from a Higher Tier paper.

5 (a) (i) What is meant by global warming? [1]

(ii) Explain why the burning of coal is partly responsible for global warming. [3]

Revision question: How is the problem caused?

(iii) What are the likely consequences of global warming? You should refer to named examples and develop the points which you make. [7]

Revision question: How does it affect people and the environment?

You need examples of those effects and must develop the ideas as there is a large mark allocation for this section. Note that this question does not ask about reducing or managing the problem.

The following answer shows how revision can result in a high mark.

Candidate's answer

Likely consequences are an overall increase in sea levels which will dramatically affect places like Bangladesh and the Netherlands. Catastrophic flooding will occur in these places. Ice and snow on mountains in the Himalayas, Alps and Pyrenees will melt. This will cause local flooding as rivers will not be able to cope with the extra water.

Bangladesh will again be a victim of this problem. Also with a higher overall temperature, hot countries such as Spain will suffer more drought and a decrease in the yield of crops. In some African countries this could cause desertification in places that once sustained life. Not all effects of global warming are negative. The south of England will experience a Mediterranean climate and it will be possible to grow grapes and other Mediterranean crops.

When studying and revising these global concepts, be careful not to mix them up. Candidates sometimes become confused between acid rain, global warming and ozone layer depletion. They are three different concepts with different causes and effects.

Of course, environmental issues are not always tested on a global scale. One topic in the syllabus deals with water pollution and this can be studied at any scale from a local stream to a large sea area.

The next question is again taken from the Higher Tier paper.

6 (a) Study Fig. 2

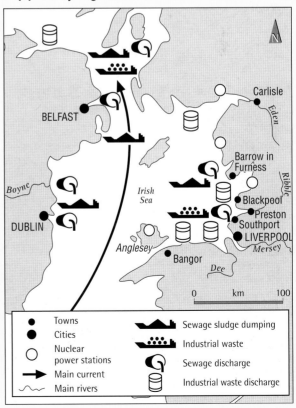

(i) Describe and explain the distribution of the different types of pollution in the Irish Sea. [4]

(ii) Explain why it is difficult for some pollutants to escape from the Irish Sea. [2]

The question begins by testing your geographical skills and understanding of **distribution**. You will meet this word in many questions so check its meaning in the glossary (page 239). Both parts of the question focus on pollution in the Irish Sea, but you may have only studied the Mediterranean Sea. However, do not be put off because the question can be answered by using the information on the map and applying your general understanding of the causes and effects of water pollution.

This question goes on to test your case study knowledge. The focus varies between the Foundation and Higher Tier papers as shown below.

Foundation Tier question

7 (a) (i) Name a river, lake or sea you have studied which has already been polluted. [1]

(ii) Describe how the pollution was caused. [3]

(iii) Describe the effects of this pollution on people. [3]

Higher Tier question

8 (a) For a named river, lake or sea, which you have studied, describe the effects of water pollution. You should refer to the effects on both people and the natural environment. [7]

Throughout the course, but especially in Unit 4, you will study topics within which people have different views on environmental issues. These different attitudes may affect the decisions which people then make about how they use or protect the environment. The following examination questions test your understanding of the different values and attitudes of people.

The first question is about how people have different opinions about mining.

Study Fig. 3

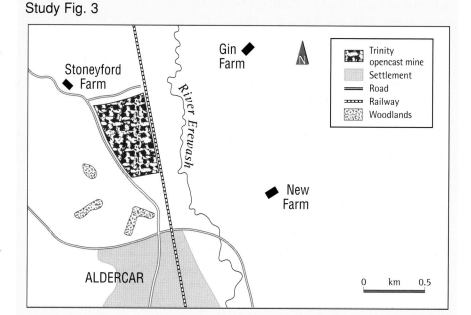

The Foundation Tier question instructs candidates to consider the problems caused by mining.

9 (a) (i) Describe different problems which the Trinity Opencast Coal Mine may cause for local farmers and residents of Aldercar. [4]

One candidate suggested these problems:

Candidate's answer

Local farmers: Noise from the mine may scare animals. Local wildlife which may help pest control may also be scared off. Any land near the mine would be showered by dust when the coal is dug up. So any crops on that land may produce a lower yield.

Residents : The land doesn't look very attractive and may cause people to move. Land values may also fall. Lorries and trains which transport the coal will cause air and noise pollution.

Examiner's comment

This is an excellent answer which shows good understanding of how people will think about the mine differently. Three problems are suggested for each group of people.

On the Higher Tier paper, the question is more wide-ranging and requires candidates to consider both the problems and benefits.

> **10 (a)** Consider the attitudes of local people to the Trinity Opencast Coal Mine. What benefits might it bring and what problems might it cause? [7]

The following answer shows the different attitudes suggested by one candidate.

Candidate's answer

The mine may bring jobs so unemployed residents of Aldercar can find employment at the mine. Many of the jobs will not require particular skills or qualifications and so can be performed by local people. This in turn improves the overall economy as more money is spent by locals so local businesses earn more. The company running the mine may spend money to improve the roads around the mine so that they can withstand the weight of heavy vehicles going to and from it. However, many people may take the opposite attitude and focus on the disadvantages. Firstly, an opencast mine spoils the once picturesque scenery and would also increase the pollution in the area – both noise pollution from the machinery, and gaseous fumes given off by the machinery. Another problem may be that lorries carrying the coal may cause congestion in the town, and due to their size may be seen as a threat to the wildlife and humans. They also give off exhaust fumes which pollute the air. Around the site of the mine are situated a number of farms. This indicates that the area has good fertile farming

land which would be removed when the mine was built. Although the soil is replaced it may never be as fertile. The location of the mine would reduce property values in Aldercar where locals would have difficulty selling their homes.

Examiner's comment

This is another excellent answer which considers both sides of the argument. The values and attitudes which local people may have are well developed. The answer was worth more than the maximum marks available for this question.

The different attitudes which people may have about the environment are also examined in Papers 3 and 4. The following question refers to the map on page 230 and is from the Foundation Tier paper.

> **11 (a)** What benefits and problems might the quarry and works cause for people living in the settlements shown on the OS map? [5]

Unlike the previous question which used a location map as a resource this question instructs candidates to examine a photograph (page 237) and an OS map extract in order to get some guidance about likely problems and benefits. This question is testing candidates' **understanding** of the attitudes of local people.

The final question is common to both Foundation and Higher Tier papers and examines the attitudes of people to oil and gas exploration out at sea. To illustrate one group's attitude to this issue the photograph which accompanies this question focuses on the protests of the Greenpeace organisation.

four

12 (a) Exploration for oil and gas is now taking place in areas of sea and ocean. Why do some people argue against searching for more oil and gas? [4]

Notice that the question only requires candidates to consider one point of view. Nevertheless there are a large number of possible ideas which can be used about both exploration and use of these energy sources. Many different reasons were suggested; below is a selection of the arguments used by candidates.

Extracts from candidates' answers

▸ *These fields could leak which would be deadly for the wildlife living in the area and could cause* *great environmental pollution in the area.*

▸ *If it spills out of tankers during transport it can kill birds, fish and other sea animals.*

▸ *Burning these fossil fuels releases carbon dioxide which causes problems such as global warming.*

▸ *We are using too much fossil fuels and we should be looking towards using alternative energy.*

▸ *People think that cleaner forms of energy ought to be exploited first.*

▸ *Instead of spending lots of money drilling for oil and gas it could be spent on developing cleaner and safer sources of energy such as solar, hydro-electric and wind power.*

TASKS

1   The following questions all focus on people and the environment. Decide whether each question is about the cause, the effects or the management of pollution.

(a) Explain how human activity interferes with the hydrological cycle causing acid rain. [4]

(b) Suggest two ways to solve the problem of acid rain [2]

(c) Why might pollution in the Mediterranean Sea be dangerous to its wildlife and the people who live around it? [2]

(d) Energy production and distribution may cause hazards to the environment. For an example which you have studied, describe these hazards and any efforts made to reduce them. [7]

(e) Explain why the level of pollution will be greater in some parts of the North Sea than in others. [4]

(f) Polluted rivers are a major problem in Britain. Suggest how water quality could be improved. Why is it difficult for people to clean up rivers? [5]

(g) Modern farming methods may cause river pollution. For an example which you have studied describe how the pollution occurs and the effects on a river ecosystem. [6]

2   The following topics may all require a case study in the examination. Check the location which you could use as a case study example.

(a) How raw material extraction in an MEDC (not the UK) affects local people.

(b) How energy production can affect the environment in an LEDC.

(c) Land use conflict in a National Park of the UK.

(d) Causes and effects of pollution of a river, lake or sea.

(e) Causes, effects and management of the problem of acid rain.

3   Questions on water pollution, especially in a sea, can provide the opportunity for a question which focuses on distribution. Answer the following question which tests map interpretation skills and understanding.

TASKS

(a) Study Fig. 5 below which shows the main sources of North Sea pollution.

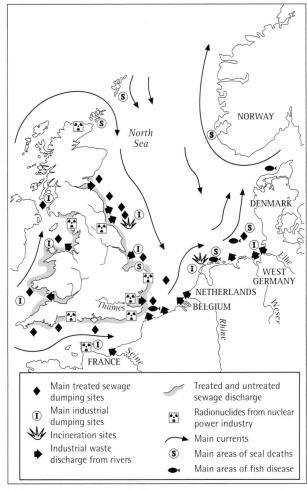

◆	Main treated sewage dumping sites	⌒	Treated and untreated sewage discharge
ⓘ	Main industrial dumping sites	☢	Radionuclides from nuclear power industry
✹	Incineration sites	➘	Main currents
➤	Industrial waste discharge from rivers	Ⓢ	Main areas of seal deaths
		🐟	Main areas of fish disease

(i) Describe the location of the parts of the North Sea which you think will be most polluted. [2]

(ii) Explain why the levels of pollutants will be greater in some parts of the North Sea than in others. [4]

4 What do the following questions have in common?

(a) Many people are worried about pollution in the North Sea. Give some reasons why the following people would want the amounts of pollution in the North Sea to be reduced:

　i　– people who own a hotel in a holiday resort on the North Sea coast [2]

　ii　– the Society for the Protection of Birds [2]

(b) Read the information below about the closure of a coal mine in Blidworth, a village in Nottinghamshire.

Describe the likely impact of the closure of the mine on the community of Blidworth. [5]

PIT CLOSURE BLOW TO VILLAGE

After weeks of speculation, miners at Blidworth got the news on Monday that they had all been dreading – the pit was to close down. The decision was announced to more than 70 miners and representatives of the mining unions, on Monday morning at the village Miners' Welfare. British Coal said that in November the pit was making a loss of over £6 million and that further reviews had shown the loss to be greater than £10 million now. A spokesman for British Coal said "We accept that there are still coal reserves available but the thickness of the seams and the geological difficulties make it impossible to mine the coal at the required cost levels". British Coal have stated that generous redundancy payments will be offered and those miners wishing to remain in the industry will be offered jobs at other collieries in Nottinghamshire.

Adapted from *Mansfield and Sutton Observer* 2nd Feb.'89

(c) Study Fig. 7

Many people think that areas with landscapes like the one in the sketch should not be changed. What reasons might they give? [3]

5 Discuss, with a partner, what attitudes people might have in the three situations described in the questions.

THE ENTRY LEVEL CERTIFICATE

People and the environment

At the end of the Entry Level Certificate course, there is a Written Test. It lasts for an hour and is worth 30% of your marks. For some of the questions, you will write the answer. For others, you need to draw or shade in or finish something off. Don't forget to take your pencil and ruler with you.

Here is part of the Test. (Don't write your answers in this book!)

1 (a) Look at the map of National Parks. How many are in Wales? _____ [1]

    (b) On the map, shade in those National Parks which reach the coast. [2]

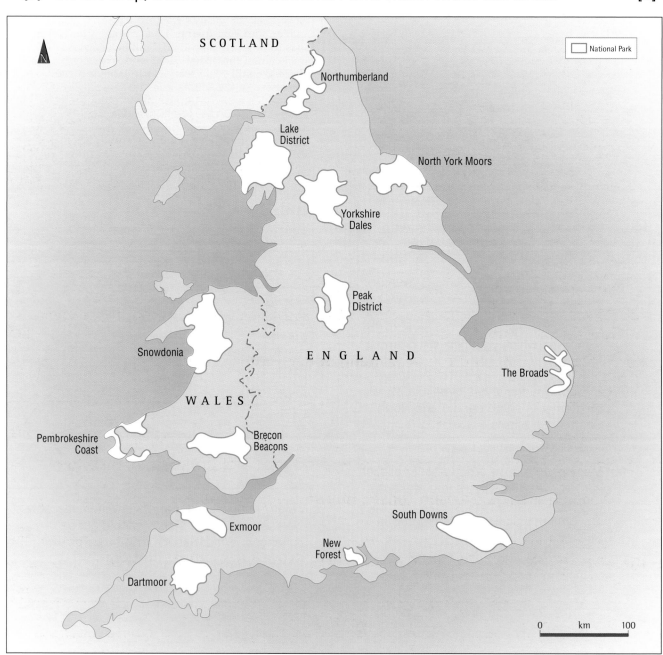

(c) Look at the graph showing where visitors to Dartmoor National Park come from.

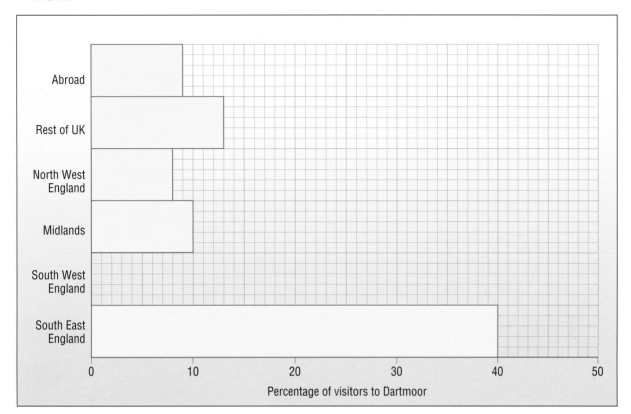

How many come from the Midlands? _____ per cent [1]

(d) 20 per cent come from the South West. Finish the graph off, by drawing the bar for the South West. [1]

(e) Write down two reasons why many people visit National Parks.

1 _____

2 _____ [2]

(f) Some people who live in National Parks think there are too many visitors.

Write about one problem caused by too many visitors. _____

_____ [2]

(g) Some people who live in National Parks want more visitors. Say why. _____

_____ [2]

_____

Advice

2 (a) Write down one of the uses of coal.

_____ [1]

Look at Photo A. A few years ago, this was an open-cast coal mine.

(b) What is now growing on this land? _____ [1]

Photo A

(c) Why are there no tall trees here? _____

_____ [2]

(d) (i) Now that mining has finished, suggest one way this land might be used.

_____

_____ [1]

(ii) Why did you choose this land use?

_____

_____ [2]

3 (a) Acid rain is a type of pollution from burning fossil fuels

Look at the diagram below.

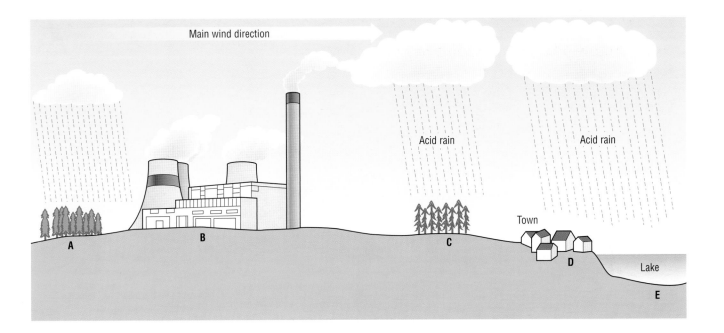

(i) What type of building is B likely to be?

_____ [1]

(ii) What has happened to the trees at C?

_____ [1]

(iii) Why has the same thing not happened to the trees at A?

_____

_____ [1]

(iv) What other effects of acid rain pollution are likely in areas D and E?

Area D _____

_____

Area E _____

_____ [2]

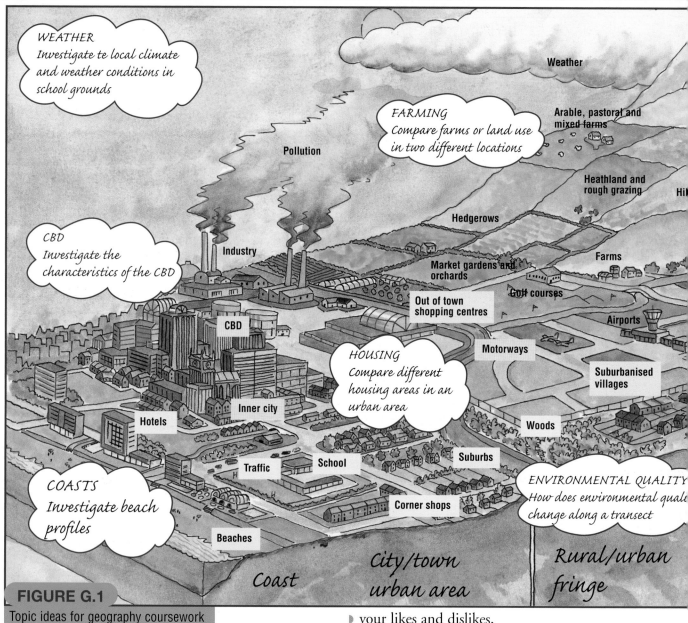

WEATHER
Investigate te local climate and weather conditions in school grounds

Weather

FARMING
Compare farms or land use in two different locations

Arable, pastoral and mixed farms

Pollution

Heathland and rough grazing

Hi

Hedgerows

CBD
Investigate the characteristics of the CBD

Industry

Market gardens and orchards

Farms

Golf courses

CBD

Out of town shopping centres

Airports

HOUSING
Compare different housing areas in an urban area

Motorways

Suburbanised villages

Hotels

Inner city

Woods

Traffic

School

Suburbs

COASTS
Investigate beach profiles

Corner shops

ENVIRONMENTAL QUALITY
How does environmental qual change along a transect

Beaches

Coast

City/town urban area

Rural/urban fringe

FIGURE G.1

Topic ideas for geography coursework

COURSEWORK IDEAS

This page suggests just a few of the many possibilities topics you can study for coursework. All are in the examination content – except one. The exception is weather. You can still undertake a topic such as investigation of the local climate and weather conditions in the school grounds, because the regulations allow you to study all topics provided that they are geographical. This is not recommended, however, because it needs extra study. There is already plenty of choice from topics that are in the content.

In practice your choice may be limited by:

▶ your likes and dislikes,
▶ the fieldwork opportunities where you live or in places you visit frequently,
▶ the need for safety while doing the work.

Once you have chosen the topic which interests you, the next stage is to decide upon title and aims. In every coursework section in this book, titles are suggested. Most titles are written down as questions. By asking a question, the focus of the work is clear and the final conclusion is easier to write. Some titles are written out as a hypothesis or as an investigation. What you should avoid is a title which begins 'A study of the River ...', which can lead to a general account only. Students who choose 'A village study' often submit coursework which is more historical than geographical.

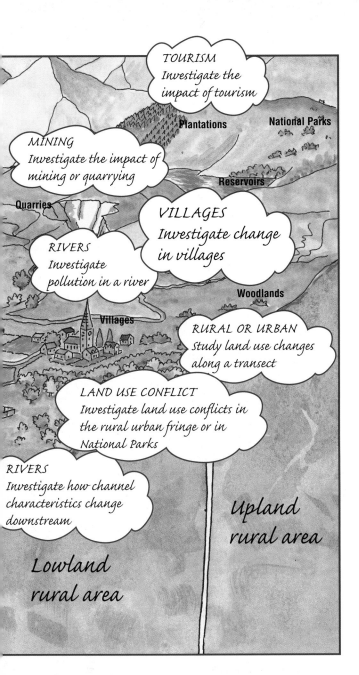

TOURISM
Investigate the impact of tourism

Plantations

National Parks

MINING
Investigate the impact of mining or quarrying

Reservoirs

Quarries

VILLAGES
Investigate change in villages

RIVERS
Investigate pollution in a river

Woodlands

Villages

RURAL OR URBAN
study land use changes along a transect

LAND USE CONFLICT
Investigate land use conflicts in the rural urban fringe or in National Parks

RIVERS
Investigate how channel characteristics change downstream

Upland rural area

Lowland rural area

THE COURSEWORK PROCESS

This can be divided up into four stages.

Stage 1 Choosing a geographical topic

▸ Think of an idea that interests you.
▸ Preferably keep it local with easy access.
▸ Talk over the idea with your teacher or with the other members of your group.
▸ Agree a working title.
▸ Write down your aims – that is what you want to find out.
▸ Read up on the geography of the topic chosen.

▸ Think about:
  – what data you need to collect
  – how you are going to collect it
  – where and when it is to be collected.

Stage 2 Collecting the data

▸ Collect as much as you can in the time available.
▸ Some must be collected by fieldwork.
▸ For your fieldwork:
  – prepare the data recording sheets and outline maps
  – assemble all the equipment needed
  – think about safety.
▸ Look for other sources of geographical information.
▸ These include maps, newspapers and books.

Stage 3 Processing and presenting the data collected

▸ Show the location of the study area on a sketch map.
▸ Label and highlight extensively any printed or copied maps used.
▸ Put in tables any statistics you have collected.
▸ Draw a variety of graphs and maps.
▸ Illustrate your work with labelled sketches and photographs.
▸ Prepare these before starting to write up and insert them where appropriate in the written work.

Stage 4 Writing it up

▸ Write about the main points shown by your maps, tables, graphs, diagrams and illustrations.
▸ Explain what they show.
▸ How do they support the title and main aims of the work?
▸ Give an overall conclusion related back to the title.
▸ Think about the geographical significance of your work.
▸ Evaluate the work done for its strengths and weaknesses.
▸ What further investigations could have been done?

USING ICT IN COURSEWORK

Using ICT for collecting data

You can use ICT to collect data as part of your fieldwork. Your school may have laptop and palmtop computers, which you can take out to record data as you collect it. If you investigate a local river, for example, measuring the width, depth and speed of the river at different locations, you can record the data straight into a spreadsheet. You can also chart the results as soon as the data has been collected. You can obtain a clearer interpretation of your results if you can see the information graphed whilst still in the area. The chart will allow you to determine if the data has been collected properly. If not, you are still in a position to consider why and repeat the tasks.

The use of digital cameras will allow you to take images in the field and, with many models, instantly view the image. If necessary, you can delete it and retake a better picture. If you can use a laptop computer in conjunction with the camera, the images can be downloaded and pasted into DTP software and annotated whilst still in the field.

Your school may have built up a collection of digital images of fieldwork locations over a number of years. The geographical skill is how you use the images in terms of labelling and integration into the text of your enquiry, rather than the taking of the photograph. A collection of images could allow you to identify change in the landscape. If, for example, the school collection of digital images included, from a field visit in a previous year, a picture of a seawall which has now collapsed, you could use both images. ICT also allows you to use the images more than once in your enquiry, such as to emphasise points in both the Results section and the Conclusion. You can also edit images, perhaps by highlighting specific points, enlarging a particular section of a picture, or by making part of an image sharper to emphasis a particular feature. It is very easy to copy and paste the images into a word processor, allowing you to make more effective use of photographs within the enquiry, integrating their use into the analysis of results.

It is also possible to use data logging equipment connected to a portable computer. If you were investigating the speed of a river, for example at different locations, you could use a flow meter.

Word processing makes coursework easier

Using a digital camera so that photographs can be downloaded in coursework

Using the Internet for coursework data

Using the scanner to create a base map

You can also collect secondary data from a CD-ROM or the Internet. A geographical enquiry about residential environments in a local community, for example, could make good use of census data from SCAMP CD-ROM for the 1991 census. You could also obtain useful data from an Internet site. For example, a local council may have a website where they promote the local area and facilities that are available. News organisations can also be useful sources of information of relevance to coursework.

Using ICT to process data

Computers can be invaluable aids when processing fieldwork data. If you produce a questionnaire in your enquiry in a shopping centre or as part of an investigation of tourism at a honeypot site in a National Park, for example, you can transfer your results into a database. Once the data has been entered, it can be quickly sorted. Elements of the data can be exported to a spreadsheet where formulae can be added to perform calculations, allowing speedy presentation and analysis of your findings.

Using ICT to present data

You can use spreadsheets and charting software to present the results of your fieldwork. This can be done very quickly, providing you with a greater amount of time for the more demanding task of analysing your results. Do not be content just to

print out the charts you produce. It would be much better to insert them into a word processor, so that you could write up your analysis of the data alongside the relevant chart. If you have used a digital camera, you can also insert the images to either word processing or DTP software ready for you to label, by using the software tools. If your school uses mapping software, you could present your results straight onto digital maps, thus making it easier for you to identify spatial patterns in your results.

Using ICT to analyse data and reach conclusions

The use of a word processor can be extremely helpful to you in producing clear, well thought out analysis and conclusions. Word processing allows you to draft and redraft your work, expanding different sections of your investigation from a simple plan or list. The order of points that you make can easily be moved around. Charts or photographs that you have produced can be inserted into the word processor, encouraging you to integrate them into your analysis. Then it is relatively straightforward to make changes to your work, when you want to improve it. Don't forget to use the spell checker and grammar checker. Do remember to save your work regularly!

COURSEWORK ASSESSMENT

A Collection and selection of data

There are 40 marks for this. The assessment scheme indicates what is expected from you to earn all these marks.

▶ Identify relevant geographical questions and issues.

This means that your topic must be carefully chosen and you must keep it geographical.

▶ Show initiative in deciding what data is required.
▶ Show initiative in recording data in an appropriate way for the aims of the work.

This means that you must be personally involved in deciding what data is to be collected and where. If you are doing teacher-led and group coursework, for which decisions about data collection have been made for you, you need to do some extra extension work of your own to show what you personally can do, if you are aiming for the higher marks.

▶ Carry out successfully the data collection.

This means that you should have collected data as accurately and carefully as possible and that you have collected sufficient data to satisfy the aims of the study.

▶ Show a thorough understanding of relevant geographical ideas.
▶ Show detailed knowledge of location and of the geographical terms related to the topic being investigated.

Advice

How to extend teacher-led and group work to ensure that marks for initiative are claimed.

1 Do the same type of data collection

  a on a different day, e.g. a shopping survey on a Saturday or market day.

  b at a different time of day, e.g. evening shoppers at an out-of-town centre.

  c at a different time of year, e.g. tourist visitors outside the main season.

  d at a different place, e.g. on another stretch of beach or in a different housing area.

2 Use a different method of data collection, e.g. an environmental survey devised by you or your own questionnaire.

This is why it is important to refer to the geographical background of the topic in your Introduction. You should use the same geographical language in your coursework as you would use in examination answers.

B Representation of data

There are 20 marks for this. They are important marks, but some students spend more time presenting their data than they do completing either of the other two sections, each of which carry double the number of marks. Not every item of data collected needs to be represented in a graph or diagram. Also there isn't one mark for every bar graph or pie chart drawn as some students seem to believe! What is more important is variety; try to include a range of techniques of presentation. Show that you have selected the best techniques for the data you have collected. If you can include among them some techniques that might be regarded as more complex, provided that they are appropriate to the data and aims of the study, so much the better. These would include proportional circles with radii calculated from square roots, isoline maps, and statistical tests such as the Spearman's Rank Correlation Coefficient (page 131). However, there is no point doing them just for the sake of doing them as they are not a requirement for full marks.

Writing up is the most difficult part of the coursework process, but you must remember that it is worth 40% of the marks. However, if you are organised and use the layout suggested below, writing up should be less painful.

Title

A clear title prominently displayed.

Table of contents

One of your last tasks after all the pages have been numbered.

Include only the major chapters.

Introduction

Explain the title and aims (purpose) of the study.

Give some information about the study area and its suitability.

Include maps of the study area.

State briefly information about the geographical background of the topic.

Data collection

Describe the methods used.

Explain why, where and when they were used.

Include examples of the types of data collected.

Analysis and interpretation of the results

Include your maps, graphs, diagrams, etc.

State what each of them shows.

Look for patterns, links, relationships, similarities and differences between them.

Comment on how they fit the title and aims.

Conclusion

Take an overall view of the work.

Summarise the work and state what it shows in relation to the original aims.

Relate your findings to their geographical background.

Briefly mention strengths and limitations.

References

References used such as books or newspapers.

People consulted.

Any other data sources such as CD-ROMs and websites.

Appendices

Examples of data collection sheets.

One or two completed questionnaires.

Tables of data.

C Analysis, interpretation and conclusions

There are 40 marks for these. They are some of the hardest marks to earn and so it is even more important to study what is expected for full marks.

▶ Write down your conclusions and discuss possible implications of these.

This means that you should try to give summary statements as you write about what the data shows and keep commenting on how this relates back to the main aim.

▶ The analysis is complete and coherent.

This means that you have used sufficient data and that the evidence has been presented in a well-ordered manner which answers the main aim(s) of the work.

▶ The work concludes with an evaluation of the validity and limitations of the evidence and conclusions.

This means that there is a final conclusion that gives an overall view of what the study has shown and there is some comment about how reliable the conclusions are. In other words, after finishing the main part of the writing up, you sit back and consider the results. Comment on what was successful and what didn't work as well as you would have wished.

▶ The student shows the ability to analyse and interpret data by applying geographical concepts and principles.

▶ The student demonstrates accurate and detailed geographical knowledge appropriate to the analysis, interpretation and conclusions of the topic being investigated.

This means that you always keep in mind the geographical background of the work. Don't be afraid to mention geographical ideas at any stage in the writing up. However, it is not an excuse for writing out chunks of textbook information. The main work involves writing about the data you have collected and what it shows.

HOW TO BE SUCCESSFUL IN THE EXAMINATIONS

Revision

You have now completed your final geography lesson. Revision is a vital part of the preparation for any examination, but how do you revise for your geography examination?

It is very important that you know what you are revising. Is it a **case study**, **geographical terminology** (the meaning of geographical words), a geographical **concept** (idea), or a geographical **skill**?

Case study

You will be allowed to choose your own case study but it must be appropriate to the question theme and taken from the correct location. As you learn a case study you should think about a question which you will be able to use your case study knowledge to answer. So, as you revise flooding in the Ganges valley in Bangladesh, for example, you may be thinking of the following question:

> Describe what has been done to control river flooding in a named LEDC.

It is *not* appropriate to use a case study of the Mississippi River to answer this question, even though you may know many river control strategies.

Questions that focus on case studies require **detailed knowledge** of a particular topic. A case study may be of a volcano or earthquake, an area of coastline, one country's population growth strategy, a city transport network, a farming system, a mining location, an area of water pollution, or many other aspects of the geography course. Therefore it is important that you can write about **one** particular volcanic eruption rather than volcanoes in general.

Geographical terminology

During your course you will have learned many new geographical words and terms. It is important that you remember the meaning of words such as subduction zone, life expectancy, fossil fuel and desertification. When revising, it is useful to compile your own **glossary** of the geographical words which you come across. A glossary is a dictionary of key words and terms with their meanings. On page 239, there is a glossary of some of the geographical words used in this book.

A glossary is useful to answer questions such as: What is meant by the term 'footloose industry'?

Geographical concepts

When revising concepts it is important that you *understand* what the idea means and how you might use it with different examples. The idea may be a:

process: such as how plate movement causes earthquakes, or how coastal erosion processes form a stack,

cause: why the birth rate of a country may change over time, or the causes of acid rain or global warming,

theory: the different characteristics of urban land use zones,

model: why the tropical rainforest ecosystem is fragile.

Once you have explained the idea and how it operates, you may then have to make use of your case study knowledge to illustrate the concept with a case study, as in the following example.

Concept: Why do many migrants into cities in LEDCs live in squatter settlements?

Case study: For a named city in an LEDC describe what is being done to improve the living conditions of people who live in squatter settlements.

Geographical skills

The best way to revise the skills that you may need to use in the examination is by practising them. There are many types of geographical skill. Some common skill questions are shown below.

OS maps

reading: (see page 230)

Use the scale [on page 233] to measure the distance between the churches in Hope and Bradwell.

Name a farm in grid square 1583.

drawing: (see page 230)

Complete the route taken by the road between Grid References 146828 and 184828.

Complete the cross-section in the frame provided and mark on your cross-section the A625 road and the railway line.

interpreting:

Describe the relief and drainage of the area north of grid line 83.

Use map evidence to explain why it is difficult to travel from Treherbert to Maerdy.

Graph

plotting:

Complete the graph by drawing the bar for caravan sites. There are 75 caravan sites.

interpreting:

Describe the changes in the amount of energy used between 1985 and 1995.

Photograph

interpreting: (see pages 230 and 237))

Use the OS map and the photograph to describe the quarry and works. You should write about site, position, and accessibility.

Satellite image

interpreting:

In which square is St Helens located?

Describe the difference in land use between the area to the north and south of the Manchester Ship Canal.

Other maps

interpreting: (see page 69)

Describe the world distribution of areas of high population density.

Data

handling:

Complete the ranking of the countries shown in the table in order of population growth per year.

Describe the relationship shown by the table between the number of people who live in the settlements and the services available.

You will not know before you open the question paper which area the OS map will cover, what type of photograph will be used, or what data a graph, such as a scatter graph, may contain. However, if you practise the skills to handle each type of resource you will be able to use unfamiliar resources.

Of course, revision should not begin after the end of the geography course. Revision is a continuous process of understanding concepts, learning the case studies which illustrate these, and practising how to apply the skills the geographer needs.

In Papers 3 and 4 the two questions may be of unequal length which may make it inappropriate to allocate time equally between the two questions. Once again, the mark allocation gives an important clue about the amount of time to spend on each section. The following formula will give a rough guide to the time allocation. In both papers it is possible to score 40 marks in 60 minutes so you should allocate approximately 1.5 minutes to each mark. By following this formula you will spend about three minutes on a 2 mark question, and about six minutes on a 4 mark question.

Remember that in the examination you do *not* have to answer the questions in the order in which they occur on the question paper. Whether you are answering in the question book (Papers 1, 3 or 4) or one separate paper (Paper 2) you may answer the questions in any order. This means that you can start by choosing a question on a subject which you know well. Hopefully by answering this first question well, it will help to dispel any examination nerves so that you can go on to attempt other more difficult questions with more confidence.

Question choice

Choosing the most appropriate question to answer is only important on Paper 2. In the other examination papers, there is no choice of questions.

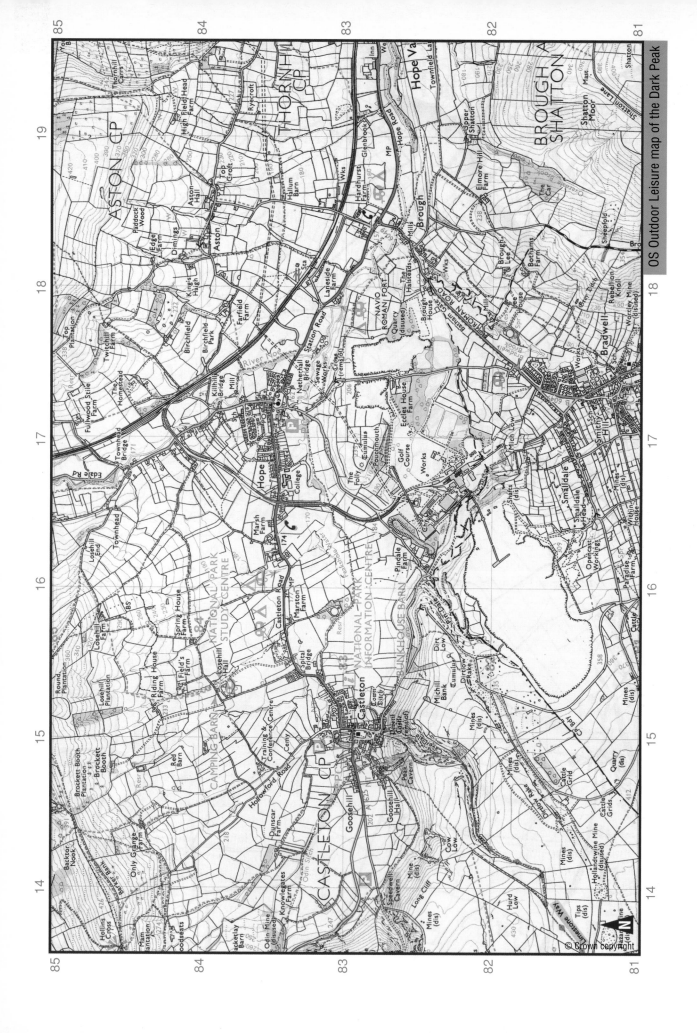

OS Outdoor Leisure map of the Dark Peak

© Crown copyright

In Paper 2, you have to choose between two alternative questions in each section. To help you to make your choice you should consider the following points:

▶ Can I answer *all* sections of the question? Do not begin to answer a question until you have read all sections of *both* questions carefully. It is poor examination technique to choose a question and begin answering the first section before you have read the remainder of the question. The first section usually has only a small mark allocation while later sections will be worth more marks.

▶ Do I *fully* understand the concept or process being tested? Some parts of the question will require a detailed explanation of the geographical ideas upon which the question is based.

▶ Do I know *in detail* an appropriate case study? This section of the question usually has the highest mark allocation and requires an extended answer.

Using your time well

All candidates answer two of the four examination papers. You will answer either Paper 1 (Foundation Tier) *or* Paper 2 (Higher Tier), and *either* Paper 3 (Foundation Tier) *or* Paper 4 (Higher Tier). Papers 1 and 2 are 2 hours long and Papers 3 and 4 are 1 hour long.

As you sit ready to begin the examination, there are so many things to think about:

▶ remember the case studies which you have learned,

▶ how to choose which questions to answer if it is Paper 2,

▶ what the different command words mean.

One other very important piece of examination technique is the appropriate use of the time available to do the examination. It is not going to help your performance in the examination if you work too slowly at first and then have to rush to complete other questions.

A simple way to plan your time is to divide up the length of time of the examination between the number of questions which you have to answer. If you are answering Paper 1 or 2, you have four questions to work through in two hours, so you should aim to spend about 30 minutes on each answer. If you budget your time in this way you will be able to give all your answers careful consideration. Remember that you can always go back to develop earlier answers or fill in any gaps if you have time at the end of the examination.

Many candidates who plan their time well by this technique find that they have a few minutes available at the end of the examination before their papers are collected. Do not waste this time. Use it to re-read your answers and use the following checklist:

▶ Have you failed to answer any section?

▶ Can you develop your ideas in any section?

▶ Have you remembered some more information about a case study?

It is always possible to improve some answers when the time pressure is reduced and you are able to review your answers carefully.

The **mark allocation** is an important clue to successful time management in the examination. The allocation is the maximum number of marks for each answer. It is shown in the examination paper by the figure in brackets ( ) at the end of each question. The mark allocation gives you a clue about how long or detailed your answer should be. Obviously the examiner expects a more detailed answer if 8 rather than 3 marks are allocated to a section. Advice to the candidate is as follows:

▶ Do not write too much detail or an over-long answer if only 2 marks are allocated to the question or the command words instruct you to describe *briefly*.

▶ A detailed answer will be required where 8 marks are allocated to a question which requires you to explain *fully* or *develop* your ideas.

THE EXAMINATION QUESTIONS

You need to examine each question carefully. Remember that they are not designed to trip you up but to give you the opportunity to show what you have learned. Your answer must be both *accurate* and *relevant*. To do this you need to be familiar with **command words** and **question themes**. Command words tell you what you have to do to answer the question. The question theme tells you what the question is about.

Command words

These can be divided into six types:

1 Simple

Name   Give   State   List   Identify

These are simple command words which provide a clear instruction. They usually indicate that a concise, brief answer is needed.

For example:

Name a fossil fuel.

Give three reasons for the increase in the amount of traffic in urban areas.

Identify the differences between a commercial farming system and a subsistence farming system.

2 Definition

Define   What is meant by   Give the meaning of

These command words are asking for definitions. The words which you will need to define will be taken from the subject content. You will have seen these words highlighted throughout this book.

For example:

Define the term **'tertiary industry'**.

What is **'sustainable development'**?

3 Description

Describe   What

Describe is the most commonly used command word. It instructs you to write about what is at a particular location, or the appearance of something, or what is shown in a resource such as a map, graph or photograph. The amount of detail required in your description is indicated by the mark allocation.

For example:

Using an example from an LEDC **describe** how energy production has affected the natural environment.

What benefits and problems might the quarry and works cause for people living in the surrounding settlements?

4 Explanation

Explain   Why   Suggest   Give reasons for

These questions are testing geographical knowledge and understanding. You are being told to account for the formation or occurrence of physical and human features.

For example:

Explain how a stack is formed.

Why do many people continue to live in cities at risk from earthquakes?

Suggest how the new industrial development may affect the local environment.

Give two reasons for the difference in discharge after the two storms.

5 Identifying differences

Compare   Contrast

These command words instruct you to write about the similarities and differences between two pieces of information.

For example:

Contrast the location of the oilfields and gasfields in the North Sea.

Compare the ways that the dependent population is supported in countries of contrasting economic development.

The best way to answer such questions is by using phrases such as 'in contrast' or 'whereas' or comparative words to bring out the difference as follows:

The gasfields are **nearer** to the coast than the oilfields.

The gasfields are located near to the English coast **whereas** (or **but**) the oilfields are located off the coast of Scotland.

It is possible to contrast by two separate statements but this is not recommended. It is better to link them with *whereas*.

For example:

> In the UK older people receive a pension. In many LEDCs older people are supported by their family.

6 Judgement

> Evaluate   To what extent

These command words are often used in more difficult questions and require you to make a judgement about a statement used in the question.

For example:

> **To what extent** are oil and gas terminals located close to the oil and gas fields in the North Sea?

The phrase 'to what extent' suggests that there is not a perfect relationship between the two pieces of information provided. Your task is to judge the strength of the link between the two variables and then to suggest where the relationship is spoiled.

While on the subject of command words, compare these two instructions.

> **Briefly** contrast the birth rate and life expectancy of the two countries described by the data. [6]
>
> Suggest reasons why the importance of different types of energy may change by 2005. You should **develop the points** which you make. [6]

These instructions give important guidance about the amount of detail required in an answer. When you look at the mark allocation, these two clues indicate how much time you should spend on answering the question.

Finally on the subject of command words, examine the four instructions below.

> You should refer to one or more areas you have studied.
>
> For a named volcano ...
>
> You may refer to examples which you have studied.
>
> For a named and located example ...

These instructions tell you that you may include at least one **case study** in your answer.

To see if you understand the command words, try writing some possible examination questions, using the photo below.

For example, for a simple question: 'Name the landform in the centre of the photograph.'

Now try: 'definition'
               'description'
               'explanation'
               'identifying differences'
               'judgement'

Question Themes

Each examination question is based on a section of the content. A theme or topic will ensure that one part of the question leads on to the next. For example, the question below is based on the theme of traffic in urban areas.

a) Study Fig. 4a

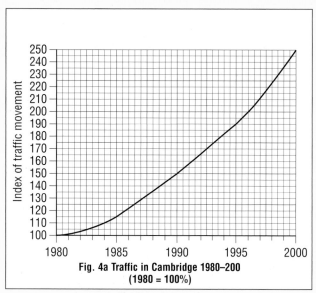

Fig. 4a Traffic in Cambridge 1980–200 (1980 = 100%)

(i) Describe what the graph shows about **traffic movement** in the city of Cambridge between 1980 and 2000.

[2]

(ii) Give **three** reasons for the **increase in the amount of traffic in urban areas**.

[3]

(iii) The **increase in traffic in urban areas may cause congestion**. What problems could result from congestion in an urban area such as Cambridge?

[5]

b) Study Fig. 4b

(i) The Metrolink scheme is part of an **'integrated network' solution to the problem of increasing traffic in urban areas**. Use evidence from the diagram to explain the term 'integrated network'.

[2]

(ii) Suggest **two possible problems of developing such a transport network**.

[2]

c) For a named town or city which you have studied, describe the strategies to **improve the movement of people and goods**.   [6]

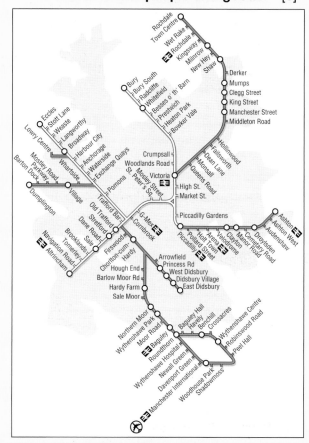

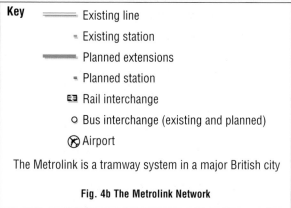

Key

══ Existing line

⹀ Existing station

══ Planned extensions

⹀ Planned station

⊟⊟ Rail interchange

○ Bus interchange (existing and planned)

⊗ Airport

The Metrolink is a tramway system in a major British city

Fig. 4b The Metrolink Network

Sometimes when a question follows a theme there may be a change in emphasis within the topic as the question progresses. The question below is based on the theme of farming but the focus moves from farming change to farming systems.

a) Study Fig. 6a

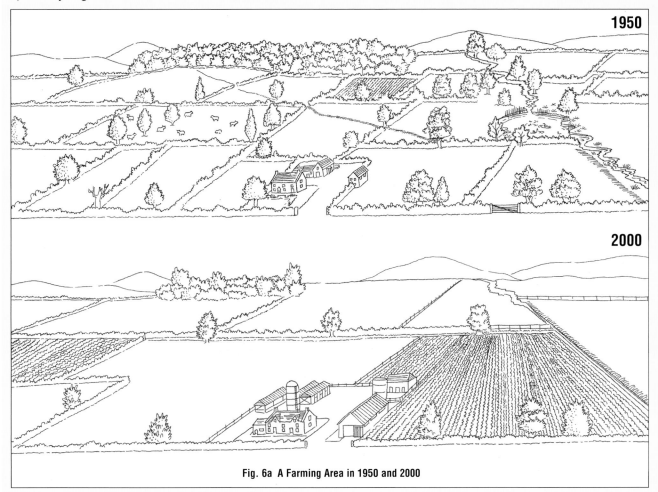

Fig. 6a A Farming Area in 1950 and 2000

(i)   Identify **two changes in the farming landscape** which have taken place between 1950 and 2000.   [2]

(ii) Explain **why the farming landscape has changed** between these two years. Do not restrict your answer only to the changes identified in part (i) of this question.   [5]

b) Study Fig. 6b

Describe how **EU policies** such as those in Fig. 6b can affect a farmer's decision making.   [4]

c) Identify the **difference between a commercial** farming system and a **subsistence** farming system.   [2]

d) For a named LEDC describe how **farming is changing**.   [7]

SET ASIDE
QUOTAS
SUBSIDIES
DIVERSIFICATION

Fig. 6b

USING QUESTION RESOURCES

A variety of resources will be used in the geography examinations. The main types of resource with which you will be familiar are:

maps: e.g. OS map, distribution map on a world scale, large-scale street plan

graphs: e.g. line graph, pie chart, bar graph, scatter graph

diagrams: e.g. sketch, process diagram, flow diagram

photographs: e.g. aerial photograph, lateral photograph, satellite image

data: e.g. table of figures

text: e.g. newspaper extract

It is unlikely that you will have seen the resource before the examination because data, photographs, maps, etc. are selected from many different places. This is not a problem, however, as the question will be testing geographical skills or understanding of the resource.

Questions which make use of resources are varied, as the following selection shows.

Select information from a graph:

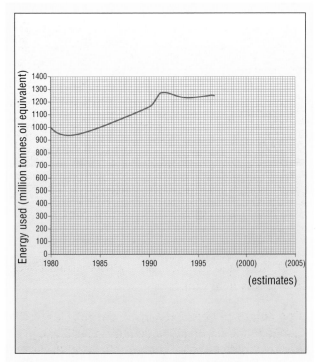

How much energy was used in 1990?

Complete a graph:

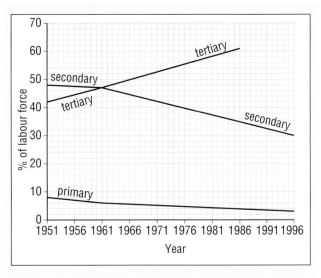

Use the following to complete the tertiary employment line.

Year	Percentage
1996	68

Complete a data table by a simple calculation:

Birth Rate (per thousand)	Death Rate (per thousand)	Population Increase
35	15	–

Questions which include a diagram, photograph or map resource usually require interpretation of the resource as shown below:

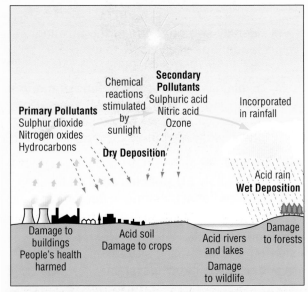

Use the ideas in Fig. 7b to describe the effects of acid rain and explain why it occurs.

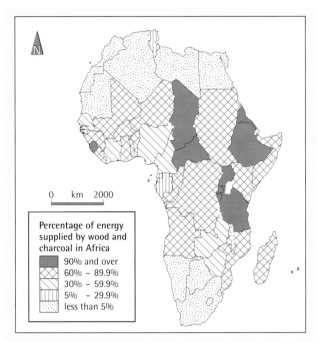

Describe the distribution of countries in Africa which depend on these forms of energy for less than 5% of their energy supplies.

OS map

This resource forms the basis of questions on Papers 3 and 4. There are many skills which must be used to interpret these maps.

You can practise them using the Landranger maps on pages 45, 103, 122 and 167. Details of the key are on page 123. Examples of maps at 1:25 000 scale are on pages 29 and 230.

A selection of these skills is outlined below.

▶ find and give 4 and 6 figure grid references,

▶ use the scale and key accurately,

▶ describe distributions,

▶ draw and annotate a cross-section,

▶ plot a routeway,

▶ relate the map to another resource,

▶ describe human and physical features,

▶ interpret the map to support geographical understanding.

Use evidence from the photograph to describe the effects of the quarry and works on the landscape and scenery.

Whatever type of resource is presented you must study it carefully as it will contain information which helps you to answer the question. Resources are not included in examination papers merely to improve their appearance. When studying maps or satellite images, the key is an important aid to understanding the resource.

The following question is an example, using pages 230, of this final skill:

Use evidence from the map to describe the benefits and problems which the quarry and works might cause for the people living in the surrounding settlements.

USING CASE STUDIES

The case study is an important section of each question in Papers 1 and 2. It is usually the final section of the question and has most marks allocated to it. Questions which refer to case studies are testing your **knowledge** of real places. The question may not always use the words 'case study'.

> Using **an example** from an LEDC, describe how energy production has affected the natural environment.

> Describe some attempts which have been made to improve the movement of people and goods in **a named town or city** which you have studied.

> For **a named river, lake or sea** which you have studied, describe the effects of water pollution on both people and the natural environment.

These questions require answers to contain details about the effects of energy production in *one* LEDC, or how the transport system has been improved in *one* town or city, or how pollution has affected *one* area of water.

Other questions allow you to use your case study knowledge in a more general answer but the question still requires details rather than vague, general statements. For example:

> Why do people continue to live in **cities** which are at risk from earthquakes? You should refer to **examples** which you have studied.

> Why do birth rates and life expectancy vary between **countries** of different levels of economic development?

The question may require your case study to be taken from a specified area, such as:

> For a named and located example of an industry **within the UK** which you have studied, explain why the industry has developed there.

> For a **named LEDC** which you have studied explain why rapid economic growth is taking place. How have the lives of people there changed as a result of this growth?

> Name a tourist area which you have studied in **the EU**. What has been done to reduce the problems caused by tourism in your named area?

When named examples are required, you will gain credit for an appropriate river or city or country, if the remainder of your answer is relevant to that example. If you cannot remember the name of your volcano case study or area of industrial change you must still describe the relevant details. You will still be given some marks for a detailed description even if the named example is missed out or incorrect.

COMMON EXAMINATION ERRORS

Finally in this chapter, some things to avoid in your geography examination. It is easy to make mistakes. Nerves, the pressure of time, trying to remember all the details of the case studies you have learned, all make it easy to make a mistake.

▶ Make sure that you follow the examination instructions in Paper 2. Do *not* answer two questions from the same section.

▶ Be careful when making your choice of case study.

Do *not* make an inappropriate choice. If the question instructs you to choose an example from an LEDC, do not choose one from an MEDC.

▶ Read the command word carefully.

Do *not* explain when only a description is required.

▶ Be aware of the mark allocation.

Do *not* write one full page of information if only two marks are allocated to a question.

▶ Be aware of how much time you are taking to answer each question or section.

Do *not* spend half the examination time on the first section and then have to rush through the remainder.

▶ Use all your time.

Do *not* sit back when you have completed your final answer. You should re-read and check your answer, adding more facts and ideas where you remember them.

▶ Use all question resources.

Do *not* ignore the map, graph or photograph. It contains information that will help you answer.

▶ Include the correct units.

Do *not* give a measurement as 2.5 when you mean 2.5 km, or 70 when you mean 70 m.

Glossary

accessible *easy to get to* 90

adult literacy rate *percentage of population able to read and write by the age of 15* 129

bar *ridge of sediment deposited across the mouth of a bay attached to land at both ends* 46

birth rate *number of live births in a year per every 1000 people* 72

braided channel *where a river channel splits into several smaller channels* 27

bus lane *part of a road not available to other vehicles to speed up bus journeys* 96

commuting *travelling daily from home to place of work* 105

condensation *water vapour changed to liquid by cooling* 24

conurbation *large urban area where towns have grown so big that they have merged* 70

convectional rainfall *rain, often from thunderstorms, resulting from warm air rising in convection currents* 182

convergent plate boundary *where an oceanic plate collides with a continental plate, resulting in the subduction of the oceanic plate* 8

core *the part of the earth surrounded by the mantle, consisting of nickle/iron compounds at very high temperature and pressure* 6

counterurbanisation *population change from urban to rural locations* 105

crust *the shell or skin of the earth, about 5km thick below the oceans but up to 50km thick under continents* 6

death rate *number of deaths in a year per every 1000 people* 72

dense *large number per square kilometre* 69

dependant population *children and old people supported by working adults* 73

discharge *the flow of water in a river [width multiplied by the depth multiplied by the velocity]* 24

distribution *where people or things are located* 214

divergent plate boundary *where plates are moving apart resulting in sea floor spreading producing new crust* 8

drainage basin *the area drained by a river and its tributaries* 24

ecosystem *a community of plants and animals, interacting with each other and their environment* 182

El Niño *warm ocean current off South America experienced every few years* 81

Enterprise Zone *area with special grants and tax advantages to anyone creating employment* 165

epicentre *point on the earth's surface directly above an earthquake's focus* 10

eutrophication *rapid growth of plants which reduce oxygen supply in water, caused by excess nutrients* 196

evaporation *water changed from liquid to vapour by heat* 24

evapotranspiration *water changed from liquid into vapour by heat or from vegetation* 24

fertility rate *number of births per year per thousand women aged 15 to 45* 72

fetch *the extent of sea, across which winds build up waves* 40

finite *resources which will eventually run out* 161

flood plain *flat land forming the floor of a valley, liable to flooding* 31

focus *underground source of an earthquake* 10

fold mountains *chain of mountains in continental crust close and parallel to a convergent plate boundary* 9

footloose *industry which has few constraints on its choice of location* 152

fossil fuels *energy sources formed over millions of years from plants and sea creatures* 160

gentrification *improving housing, including creating new, often expensive homes in rundown inner cities* 107

green belt *area of countryside around an urban area where most new buildings are not permitted* 95

greenfield *urban or industrial growth on a previously rural site* 153

greying population *high, and increasing, proportion of elderly people* 74

groundwater flow *movement of water through porous rocks* 24

groyne *breakwater fences to stop longshore drift removing beach material* 49

guestworker *immigrant to affluent country from poorer country attracted by higher paid work, health care and education* 78

high order goods *things bought less frequently, expensive* 91

honeypot *location prone to overcrowding* 194

hydrograph *graph showing precipitation and a river's discharge at a place over a period of time* 24

impermeable *rocks which do not absorb water* 45

industrial inertia *the survival of an industry in an area after its locational advantages have disappeared* 152

infant mortality rate *number of babies dying before their fifth birthday, per thousand population* 129

infiltration *water seeping down through soil* 24

infrastructure *the basic framework of power and water supplies, transport networks and services such as education and sewerage* 135

interception *when plants prevent rainfall reaching the ground* 24

kiss and ride *where the car is driven away as the passenger transfers to public transport* 97

lag time *the time between peak precipitation and peak discharge as shown on a hydrograph* 25

levée *embankment alongside a river* 30

life expectancy *how many years a new born baby can expect to live* 129

load *sediment carried by a river, in solution, suspension or rolled along its bed* 26

longshore drift *movement of sand and pebbles along a beach by waves* 46

low order goods *things bought frequently, usually cheap* 102

malnourished *inadequately balanced diet* 133

mantle *the part of the earth between the crust and the core* 6

migration *movement of people* 72

multiplier effect *the spiral of growth or decline from the creation or closure of an economic activity* 137

natural increase *the difference in a year between the number of births and deaths* 72

neighbourhood centre *group of low order shops serving a community within a town* 104

overland flow [surface flow] *movement of rainwater which does not percolate into the soil and rocks, nor is evaporated* 24

ox bow lake *meander cut off from the main river channel* 30

park and ride *providing car parks away from town centres with public transport to the CBD* 97

pedestrianisation *closing streets to traffic* 91

permeable *rocks which can absorb water* 45

plate tectonics *large scale earth movements where pieces of the crust [plates] move relative to one another* 7

precipitation *water from the atmosphere such as rain or snow* 24

prevailing wind *the most frequent wind direction* 42

raw materials *naturally occurring substances from which the making of all goods starts* 152

refugee *someone forced to move to live elsewhere* 79

relief *the shape of the land* 71

Richter scale *measure the power or energy of an earthquake* 10

runoff *water flowing across the surface into and in rivers* 24

salt marsh *area with salt-loving plants on the landward side of a spit or bar, periodically flooded by sea water* 47

saturation *when porous or permeable rock cannot hold any more water* 24

savannah *tropical grassland* 132

seismic waves *shock waves from the focus of an earthquake* 10

seismograph *instrument for measuring earthquake vibrations [seismic waves]* 10

sparse *few per square kilometre* 69

sphere of influence *area served by a settlement, shop or service* 108

spit *beach extension formed by longshore drift* 46

subduction *when an oceanic plate is forced beneath a continental plate, melting it* 9

sustainable *using resources no faster than natural processes can replenish them* 189

threshold population *minimum number of customers required to keep a business going* 100

throughflow *flow of water down a slope through the soil* 24

TNC *Trans-national companies, usually run from MEDCs with branch factories worldwide* 137

tombolo *beach joining an island with mainland* 46

transform plate boundary *where a plate is slipping past another plate, earthquakes marking when friction is overcome* 8

transpiration *the transfer of water to the atmosphere through leaves* 24

water table *the level below which porous or permeable rock is saturated with water* 24

wave refraction *the bending of a wave as it approaches the shore caused by shallowing water* 41

weathering *breakdown of rocks by natural agents such as rain, frost, ice and wind* 43

Place Index

Locations of principal case studies are shown in bold.